Berkeley Hill

The Essentials of Bandaging

With Directions for Managing Fractures and Dislocations. Fifth Edition

Berkeley Hill

The Essentials of Bandaging
With Directions for Managing Fractures and Dislocations. Fifth Edition

ISBN/EAN: 9783337811648

Printed in Europe, USA, Canada, Australia, Japan

Cover: Foto ©berggeist007 / pixelio.de

More available books at **www.hansebooks.com**

ESSENTIALS OF BANDAGING.

THE

ESSENTIALS OF BANDAGING

WITH DIRECTIONS

FOR MANAGING FRACTURES AND DISLOCATIONS;

FOR ADMINISTERING ETHER AND CHLOROFORM, AND
FOR USING OTHER SURGICAL APPARATUS;

AND

CONTAINING A CHAPTER ON SURGICAL LANDMARKS.

Illustrated by 136 Engravings on Wood.

BY

BERKELEY HILL,

M.B. Lond., F.R.C.S.

Professor of Clinical Surgery in University College, Surgeon to University College
Hospital and Surgeon to the Lock Hospital.

FIFTH EDITION.

REVISED AND ENLARGED.

NEW YORK:

J. H. VAIL & COMPANY.

1883

PREFACE TO THE FIFTH EDITION.

THIS little work has now reached a fifth edition, for which it has been carefully revised and has received several additions.

A description of Laryngoscopy has been contributed by Dr. Vivian Poore, and one of Ophthalmoscopy by Mr. John Tweedy, Professor of Ophthalmic Surgery in University College : while Mr. Bailey has revised the instructions for administering Anæsthetics. My thanks are due to these gentlemen for their kindness in thus ensuring for my readers thoroughly trustworthy information on these special subjects.

Other additions have been made ; notably, accounts of Thomas's splints, of Croft's plaster of paris splints, of Carr's splints, and of several other improvements in practical surgery, and every effort has been made to render the former text correct.

Several new drawings have been inserted into this Edition.

55, WIMPOLE STREET,
    December, 1882.

CONTENTS.

CHAPTER I.

BANDAGING.

PAGE

General rules.—Different materials of bandages.—Bandage winder.—Position of the operator.—Mode of holding the bandage.—Varieties of turns; the simple spiral; the reverse; the figure of 8 . . . . 1—4

Bandaging the head.—The common roller.—The knotted bandage.—The capelline bandage.—The shawl cap.—The four-tail bandage.—Applying cold to the head.—Compressing the jugular vein . . . . . . 4—9

Bandaging the trunk.—The breast.—The groin.—For tapping the belly.—Southey's cannulæ.—The T-bandage.—The straight jacket.—Manacles for delirious patients.—To suspend the testicles . . . . . . 9—14

Bandaging the upper extremity.—The fingers.—The thumb.—The hand.—The fore-arm.—The elbow.—The arm.—The shoulder.—The axilla.—Wound of the palmar arch.—Bleeding at the elbow 15--21

Bandaging the lower extremity.—The foot.—The leg.—The thigh.—The heel.—The toe.—The knee.—A stump.—Extending a stump.—The many-tailed bandage.—Elastic socks . . . . 21—25

CHAPTER II.

STRAPPING.

PAGE

General rules.—Strapping the breast.—The testicle.—Ulcerated legs.—Joints.—The ankle.—Scott's mercurial dressing . . . . 26—3

CHAPTER III.

TREATMENT OF FRACTURES.

The head and trunk.—Of the lower jaw; by the external splint and bandage; by interdental splints.—Of the ribs; by plaster; by a body roller.—Of the pelvis . 32—36

The upper extremity.—Of metacarpal bones; by a gutta-percha glove; by a ball of tow.—Of phalanges.—Of the lower end of the radius; by the pistol splint; by the gutta-percha gauntlet; by Carr's splints.—Of both bones of the fore-arm.—Of the olecranon; by figures of 8 and an inside splint; by a gutta-percha sheath.—Of the humerus near the elbow; by lateral hollowed splints; by a gutta-percha L-shaped splint.—Stromeyer's cushion. —Of the shaft of the humerus.—Of the anatomical or surgical neck, and of the great tuberosity of the humerus, by a cap for the shoulder; by simple confinement.—Of the acromion.— Of the clavicle; by an axillary pad and elevation of the elbow; the American ring pad; by Sayre's method; by a figure of 8 behind the back . 36—60

The lower extremity.—Rupture of the tendo Achillis.—Separation of the epiphysis of the os calcis.—Fracture of the fibula, by Dupuytren's splint; by starch or plaster case.—Of the tibia, by McIntyre's splint; slinging the splint; elevating it on a block.—Transverse fracture of

PAGE

the tibia, by lateral splints; in the flexed position; by
horse-shoe anterior splint.—Of the patella; by back
splint and figure of 8; by starch bandage.—Of the
shaft of the femur; by Liston's mode of using the long
splint; by using elastic extension; Coxeter's elastic
perineal band; elastic stirrup; Brown's stirrup; the
Scotch method; Hamilton's splint for children; by
continuous extension with the limb bent; tendency to
angular union; double incline planes; slinging the
double incline planes in fracture of the neck of the
femur; by continuous extension of weight and pulley 60—86

The starch bandage.—The plaster of paris bandage.—Croft's
plastic splints.—Gum and chalk and other stiffening
mixtures. — Sand-bags. — Sayre's plaster jackets.—
Cradles.—Salter's swing cradle; Canopy cradle   . 87—104

Leather splints; Splint for the hip.—Gooch's flexible splint.
—Poroplastic felt   .   .   .   .   .   . 104—108

CHAPTER IV.

DISLOCATIONS.

General rules.—Of the lower jaw.—Of the clavicle.—Of the
shoulder; signs of dislocation into the axilla; when beneath the clavicle; when behind the scapula. Modes of
reduction; by the heel in the axilla; the clove-hitch
knot; by simple extension.—Of the elbow, signs when
both bones go backwards; distinctions between dislocation and fracture near the elbow; the mode of reduction
by the knee inside the fore-arm; by extension below the
elbow.—Of the radius only, by extension at the wrist.—
Of the thumb and fingers; handle for commanding the
phalanx.—At the hip; signs of dislocation backwards,
reduction by extension; by manipulation or leverage;

PAGE

signs of dislocation downwards, modes of reduction ;
signs of dislocation on to the pubes, modes of reduction.
—Of the knee ; incomplete, lateral, and posterior; modes
of reduction.—Of the patella, modes of reduction.—Of
the foot, modes of reduction . . . . 109—126

Scarpa's shoes.—Points to be attended to in fitting the shoe.
—Thomas's splints for the hip and for the knee.—Casting in plaster of paris . . . . . . 126—134

CHAPTER V.

MISCELLANEOUS.

Opthalmoscopy.—The eye douche.—Syringing the ears.—
Nasal douche ; Epistaxis ; Ice-cold injection ; Plugging
the nares ; Belloc's sound.—Drawing teeth ; varieties
of forceps ; extracting incisors and canines, bicuspids,
upper molars, lower molars, wisdom molars, roots ; the
elevator.—Stopping bleeding after extraction.—Laryngoscopy.—Sore nipples ; nipple shields.—Plugging the
vagina ; Kite's tail-plug for vagina.—Injecting the
urethra.—Catheters, silver ; different kinds of flexible
catheters and bougies ; sounds.—Passing catheters and
bougies ; conformation of the urethra ; difficulties ;
passing the female catheter.—Washing out the bladder.
—Tying in catheters.—Position for lithotomy.—Brown's
tampon.—Bed-sores, application to prevent the formation of bed-sores ; the floating-bed ; the water-cushion.
—The coin-catcher.—The stomach-pump, when used to
empty the stomach or to inject food.—Washing the
stomach by a syphon.—Transfusion of blood ; Graily
Hewitt's method ; Roussel's method. — Tourniquets ;
Petit's ; make-shift ; Bloodless operations ; Elastic tourniquets ; Signoropi's ; Lister's ; Davy's lever ; Carte's.

PAGE

—Mercurial fumigation ; general ; local.—Hot air baths.
—Vapour baths.—The aspirator.—Cupping.—Leeches.
—Stopping leech bites.—Tents.—Setons.—Drainage
tubes.—Issues.—Trusses, requirements of, inguinal,
femoral, umbilical, Salmon and Ody's.—Canteries, iron ;
galvanic ; Paquelin's.—Caustics.—Vesicants, mustard,
cantharides, iodine.—Corrigan's hammer.—Poultices
and fomentations.—Lister's antiseptic method ; chloride
of zinc ; Volkmann's spoon.—Boracic acid ; salicylic
acid ; Iodoform.—Irrigation ; spiral coil.—Leiter's
tubes.—Esmarch's irrigator.—Administration of chloroform, precautions ; dangers ; methods.—Of ether ;
Clover's inhaler.—Ethidene dichloride ; Bichloride of
methylene ; Nitrous oxyde gas.—Artificial respiration ;
Marshall Hall's, Silvester's, and Howard's methods.—
Local anæsthesia ; ether spray ; pulverised fluids ; chloroform vapour to the uterus.—Subcutaneous injection.—
Collodion.—Vaccination .   .   .   .   .   135—240

CHAPTER VI.

SURFACE-GUIDES AND LANDMARKS.

The Head, cranial region ; facial region ; cavity of the nose ;
of the mouth.—The Neck, anterior region ; lateral
region ; posterior region.—The Thorax, the front ; the
heart ; the lungs ; the great vessels ; attachment of the
diaphragm ; The side, region for tapping ; The back,
origin of the spinal nerves ; situation of the great
plexuses ; distribution of main cords.—The Abdomen ;
position for examining.   Surface marks in front ; Linea
alba, parts it overlies ; position of gravid uterus at different periods ; Operations performed in linea alba.—
Regions of the abdomen ; arrangement of viscera therein ; abdominal aorta ; its chief branches.—Inguinal

PAGE

hernia; landmarks of scrotal tumours.—Femoral hernia.
—Employment of taxis.—Fold of the groin.—The Perinæum ; limits, contents.—Examining the rectum ; parts
to be felt within the anus.—The Upper Extremity, the
shoulder surface-marks ; the alteration of shape produced by injuries ; mode of examination.—The Arm.—
The elbow ; bony landmarks ; hollow in front of the
elbow ; alterations of shape through injury.—The Forearm.—The Wrist ; landmarks.—The back of the Hand ;
the Palm ; the Fingers.—The Lower Extremity.—The
Groin, the Hip, the Buttock : their bony landmarks,
Dislocation and Fracture ; Nelaton's line ; Bryant's
rule ; change in shape of soft parts from disease.—The
Femoral Artery.—The gluteal, the ischiatic, the pudic
arteries.—The knee; the patella; the ham; the tumours
of the ham.—The leg; the calf.—The ankle.—The foot;
landmarks of the foot   .   .   .   .   . 241—308

APPENDIX.

LIST OF APPLIANCES FOR THE OPERATING ROOM AND THE
    SICK ROOM.—LIST OF SEDATIVES AND RESTORATIVES.
    —FOR THE ARREST OF HÆMORRHAGE .   .   . 309—313

LIST OF INSTRUMENTS EMPLOYED IN OPERATIONS—
ON THE HEAD AND NECK.

Trephining the skull.—Operations on the eye.—Hare-lip.—
    Resection of the jaw.—Excision of the tongue.—Cleft
    palate.—Excision of tonsils.—Laryngotomy.—Tracheo-
    tomy .   .   .   . 314—318

On the Trunk.

PAGE

Removal of breast or tumours.—Nævus.—Tapping the pleura.
—Tapping the belly.—Colotomy.—Ovariotomy.—Cæsarian section.—Strangulated hernia.—Radical cure of
hernia.—Hæmorrhoids.—Fistula in ano.—Cleft perinæum.—Extirpation of the cervix uteri.—Amputation
of the penis.—Circumcision.—Excision of testis.—Tapping a hydrocele.—Vesico-vaginal fistula.—Retention of
urine.—External urethrotomy. — Lithotomy. — Lithotrity.—For removing foreign bodies from the urethra
and bladder . . . . 319—327

On the Limbs.

Ligature of the larger arteries.—Resections.—Removal of
necrosed bone.—Amputations . . . 328—331

INDEX 333

ESSENTIALS OF BANDAGING,

&c.

CHAPTER I.

BANDAGING.

GENERAL RULES.—Ordinary bandages are strips of unbleached calico 6 or 8 yards long, having a breadth of ¾ inch for the fingers and toes, 2, or 2¼ inches for the upper limb, 3 inches for the lower limb, and 6 inches for the body. These, when tightly rolled for use, are termed rollers. Besides these rollers for general use there are special bandages, such as rollers of muslin for using with plaster of Paris, of stocking webbing when great softness is desired, of welsh or domett flannel for irritable skins; or, of india-rubber where elastic tension is needed.

The annexed figure on page 2 depicts a small winder for rolling bandages, invented by the late Mr. Clover, and improved by Mr. Coxeter.

Position of the Operator.—He should place himself opposite his patient, not at the side of the limb to be

B

bandaged ; the limb too should be bent to the position it will occupy when the bandage is completed.

Before applying any kind of apparatus, the surgeon should see that the limb is carefully washed and dried.

How to hold a Roller.—When applying a roller it is best to begin by placing the outer surface of the

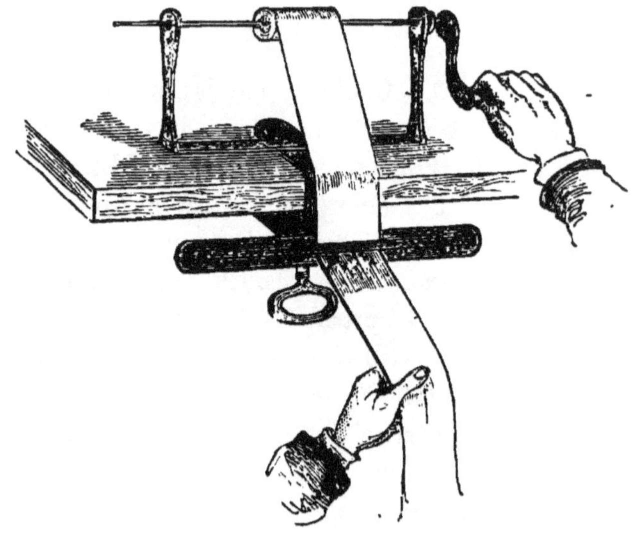

Fig. 1.—Bandage winder.

roller next the skin (see fig. 3, page 4), for it then unwinds more readily, and the first turns are more easily secured ; moreover the bandage should be carried from the inner side of the limb *by the front* to the outer side, for the muscles are thus more firmly and pleasantly confined than by turns passing in the opposite direction ; of course this observation supposes the patient's hand and forearm to be in their usual position with the thumb upwards.

Varieties of Turns.—In carrying a bandage up a limb, it is necessary, in order to support the parts evenly, to employ a combination of three different turns. The *simple spiral*, *reverse*, and the *figure of* 8.

The *simple spiral* turn is used only where the circumference of the part increases slightly, as at the wrist ; but when the limb enlarges too fast to allow the fresh turn to overlap the previous one regularly, the turn must be interrupted, and the bandage brought back again by reverse, or by figure of 8.

To *reverse* the bandage, the thumb of the unoccupied hand fixes the lower border of the bandage at

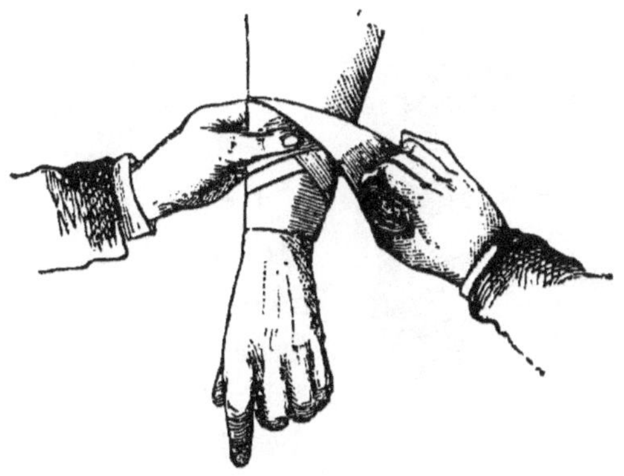

Fig. 2.—Reversing a roller.

the highest point of the turn while the roller is turned over in the other hand, and then passed downwards to overlap the previous turn evenly. At the moment of reversing (see fig. 2), the bandage should be held *quite slack,* and not unrolled more than is necessary to make the reverse. All the reverses must be carried one above the other along the outer side of the limb, and only employed where really necessary.

Figures of 8 are made, as their name implies, by passing the roller alternately upwards and downwards as it enwraps the limb (see fig. 3). They are adopted where the enlargement is too great and irregular for reverses to lie evenly, over the ankle and elbow for instance.

Fig. 3.—Figure of 8 turn.

THE HEAD.

Bandages for the Head.—A roller is commonly applied in three different ways to the head. 1st. For keeping simple dressings in place.

Apparatus.—1. A roller 2 inches wide, and of the usual length. 2. Some pins.

A turn is first carried horizontally round the head above the brows and below the occipital protuberance

Fig. 4.—Bandage for retaining dressings in position, showing two sets of oblique turns.

and fastened by a pin ; this being done, the roller is carried across the dressing, and getting into the line of the first turn, is passed round the head again, then across the dressing and round the head by oblique and horizontal turns alternately, the latter to fix the former (see fig. 4). In the figure the oblique turns have been doubled, and would fix dressings on each side of the head.

The shawl cap and fourtail bandage are also used to keep dressings in place (see p. 7.)

Knotted Bandage.—This is used when pressure on the superficial temporal artery is required.

Apparatus.—1. A bandage 8 yards long, 2 inches wide, rolled into 2 heads. 2. Some lint. 3. A piece of a wine cork one-third of an inch thick. 4. Needle, thread, and pins.

The cork is folded in a double thickness of lint; over this are placed six or eight more folds of lint of gradually increasing size, and the whole are kept in shape by a stitch passed through them and through

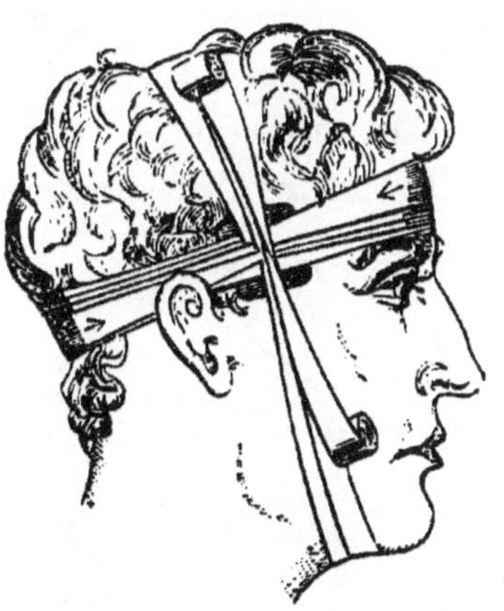

Fig. 5.—Knotted Bandage.

the cork. When the corners are trimmed away this forms a *graduated compress*, and is laid on the wound with the small end downwards.

One end of the roller is taken in each hand, and the middle laid over the compress on the injured temple, say the right ; the ends are next carried round the head, one just above the eyebrows, and the other backwards below the occipital protuberance, till they meet at the left temple, where they are passed from one hand to the other to be brought to the wounded temple. Here they are again tightly crossed, one end being carried under the chin and by the left side to the vertex, there meeting the other end, which has passed over the head in the opposite direction (see fig. 5). Here the hands again change ends, and the bandaging is continued till each end reaches the right temple. There they are again crossed or "knotted," but this time they are passed horizontally round the head. Having done this the ends are pinned and cut off, or if necessary the knots are repeated before fastening; but the first pair, if tightly drawn, suffice as well as several.

To insure firm pressure, care should be taken that each "knot" overlies its predecessor.

The Capelline Bandage is rarely required, but is used when the restlessness of the patient renders it difficult to keep dressings or ice-bags in their place.

It is also the bandage employed to keep the dressings on the stump of a limb.

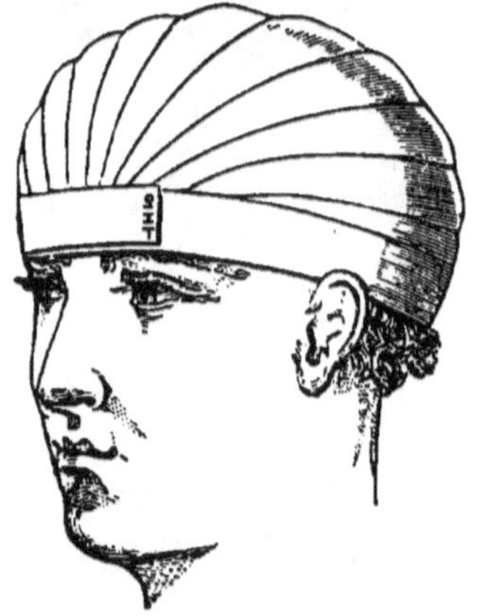

Fig. 6.—Capelline Bandage.

Apparatus.—1. A double-headed roller, 2 inches wide and 12 yards long. 2. Some pins.

The middle of the roller is laid against the forehead just above the brows, and the ends passed behind the occiput, where they are crossed. After this, one end continues to encircle the head horizontally, fastening down at the forehead and occiput alternately the other end of the roller as it goes forwards and backwards. The second head of the bandage, starting from the occiput, is brought over the top along the middle to the bridge of the nose, and passes under the encircling turn, which fixes it. It is then carried back to the occiput, on the right of the mesial band; when again fixed behind by the encircling turn, it is brought forward on the left side of the mesial band and fixed in front. This arrangement is repeated until the head is covered in a closely fitting cap (see fig. 6).

In beginning this bandage, it is necessary to keep the first circle low down, close to the brows in front, and below the occipital protuberance behind, or the cap will not fit firmly over the skull.

A Shawl Cap is readily improvised with a silk or cambric handkerchief folded diagonally into a triangle; the base of the triangle is then carried over the brow, the apex let fall behind the occiput, where the ends cross, and catching in the apex, come round to the front to be tied on the forehead.

Fig. 7.—Shawl Cap.

The shawl cap is readily applied to retain dressings on a stump.

The Four-tail Bandage.—Instead of applying the handkerchief in the manner just described, it may be split from each end to within six inches of the middle, and so converted into a broad four-tail bandage; the middle is laid on the top of the head, the hinder ends tied under the chin, and the forward ones behind the nape of the neck. Or a piece of calico, $1\frac{1}{2}$ yards long and 6 inches wide, is split from each end to 3 inches short of the centre—one pair of tails being rather wider than the other. If used on the face, the middle is put against the point of the chin, the two narrow tails are carried backwards to the nape, crossed, and pinned together on the forehead above the brows. The two broader tails are carried upwards in front of the ears, where they turn round the two narrow tails, to be either tied or pinned at the vertex. Four-tail bandages are used elsewhere, as in the axilla, to keep poultices in place, &c.

To apply Ice-Bladders to the Head.—This is done by folding a thin napkin over the bladder, which is then laid against the head or part to be kept cool, and the ends of the napkin are pinned tightly down to the pillow at each side. Fastened in this way the bag cannot slip, and its weight is at the same time prevented from pressing on the head. An india-rubber skull-cap, attached by a band under the chin, through which ice-cold water circulates, is used for this purpose. Again, soft metal gas-pipe tubes, called Leiter's tubes, are rolled into concentric rings or spiral coils, and bent to fit the head or other part to be cooled.

These coils are fastened on with straps, and cold water is allowed to flow continuously through them. If the application of heat rather than its abstraction be desired, warm water can be equally well directed through the coil.

To compress the Jugular Vein after bleeding. —After venæsection of the external jugular vein it is requisite to keep a compress of lint on the wound. This is done by fastening a ¾-inch-wide bandage on the neck with two simple turns ; it is then carried in a figure of 8 round the neck, across the compress and under the armpit of the further side, and round the neck again. If the figure of 8 is passed pretty firmly, sufficient pressure is made to check bleeding without interfering with the circulation through the vessels ; the turns round the neck of course must not be tight.

THE TRUNK.

To bandage the Breast.

Apparatus.—1. A roller 3 inches wide and 8 yards long. 2. Pins.

The roller is first car- ried once round the waist below the breast, be- ginning in front and passing towards the *sound* side. When the bandage is fixed, the roller ascends over the lower part of the diseased breast to the opposite shoulder, and comes back by the armpit to the

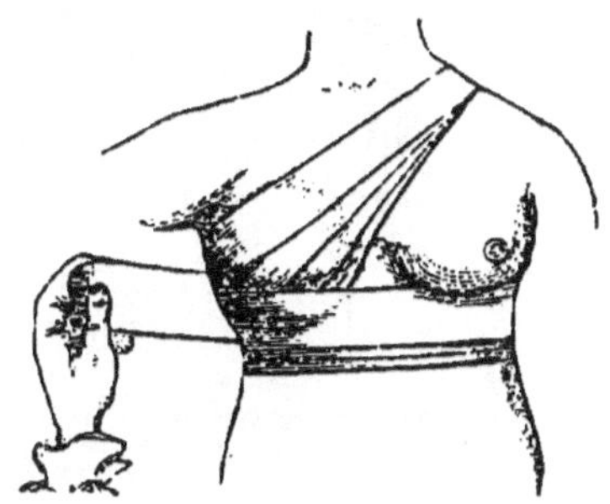

Fig. 8.—Bandage for a Breast.

horizontal turn ; where it passes round the waist to fix the oblique turn. The bandage is continued by repeating the oblique turns over the breast and shoulder, and the horizontal turns round the body until the breast is fully compressed. Each turn over the breast is carried higher than the preceding one, and each turn round the body overlaps the oblique one to keep it in place (see fig. 8).

To bandage both Breasts.—This is readily done by first bandaging one breast in the manner described: then, having carried the roller over the shoulder of the side already bandaged, bring it across the sternum and under the second breast to the horizontal turns, which it follows alternately with oblique ones, as was done in bandaging the first breast. The only difference is that, in compressing the first breast the bandage was passed obliquely ' upwards, for the second it is carried obliquely downwards over the breast.

Spica Bandage at the Groin.

Apparatus.—1. A roller $2\frac{1}{2}$ or 3 inches wide. 2. Some pins.

Lay the end on the groin to be bandaged, carry the roller between the top of the thigh bone, (great trochanter,) and the crest of the hip behind

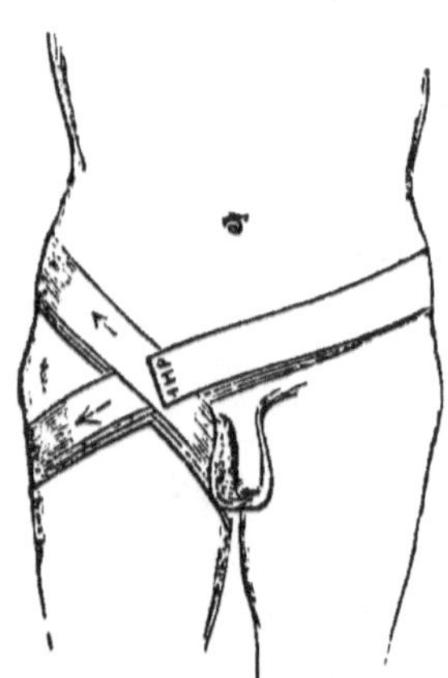

Fig. 9.—Spica for the Groin.

the pelvis to the other side of the body, there passing also between the crest of the hip and trochanter. Next take the roller downwards in front of the pubes and the injured groin, outwards and round the thigh *below* the trochanter to the fold of the buttock, and pass it up between the thighs to the groin, where the figure of 8 is completed. More similar turns are to be passed in the same way round the body and below the buttock (see fig. 9). At the groin each turn should overlap, but lie a little above the preceding turn. A pin, when the necessary number of turns is completed, fastens down the end.

Body bandage for tapping the Belly of Ascites. —This is made of two thicknesses of stout flannel, 2 feet wide in the middle, where it forms a continuous sheet for 18 inches, but beyond that it is split into 3 tails, 6 inches wide and 3 feet long. In the middle line, 4 inches below the centre, is a round hole 2 inches across, through which the surgeon reaches the skin to insert the trocar. The trocar generally employed has a diameter of nearly $\frac{1}{2}$ an inch, and on its cannula a tap is fitted in order that, as the trocar is withdrawn after puncture to make way for the flow of fluid, the escape of fluid may be prevented until a length of india-rubber tubing is attached to the trocar to conduct it into a bucket placed near.

When in use, the middle of the bandage is placed in front with the hole in the mesial line of the body, and midway between the umbilicus and pubes; the ends of the right side are passed behind the back to the left, interlacing with those from the left side. When all is ready, an assistant standing on each side of the bed pulls steadily on the ends to keep up continuous pressure on the abdominal viscera as the fluid escapes. After the fluid is evacuated the ends are wound firmly round the body in front, while the puncture in the wall of the belly is closed by a fold of lint attached with a strip of plaster, or to prevent the troublesome oozing of fluid which sometimes follows the operation, a hare-lip pin may be inserted, and the edges thus retained in apposition by a twisted silk suture.

Southey's Cannulæ.—Dr. Southey employs fine cannulæ to insert into the cellular tissue of dropsical limbs to drain away serum. They are silver tubes with sharp points, 3 inches long, one millimetre in diameter, and fitted with fine india-rubber tubing to carry off the fluid as it trickles out of the cannulæ into a receptacle. In cases of extreme debility, where ordinary tapping is not available, ascitic fluid may be drawn from the belly by their means.

The T Bandage is used to apply dressings, compresses, &c., to the anus or perinæum. A roller 3 inches wide is fastened by a couple of turns round the pelvis, and then fixed by a pin at the middle line in front. From this point the roller is carried tightly over the dressings to the corresponding point behind, and returned once or twice more until sufficient pressure is gained, when it is fastened off. The bandage may be prepared beforehand by sewing to the centre of a strip of calico 3 inches wide and 1½ yards long, a similar strip a yard long in the shape of the letter T. When applied, the longer strip is carried round the body with the attached piece at the sacrum, and made fast in front. The attached piece is then brought tightly forward and fastened also in front.

The Strait Jacket is made of jean or stout canvas. It is cut long enough to reach below the waist, around which a strong tape is carried to be drawn tight and tied after the jacket is put on. The sleeves are several inches longer than the arms, and their ends can be drawn close by a long tape which runs in the gathers; a similar tape confines the garment round the neck, and it is tied behind by tapes attached down the sides. When the jacket is to be put on a patient, it is first turned inside out, then one of the nurses or assistants thrusts his own arms through the

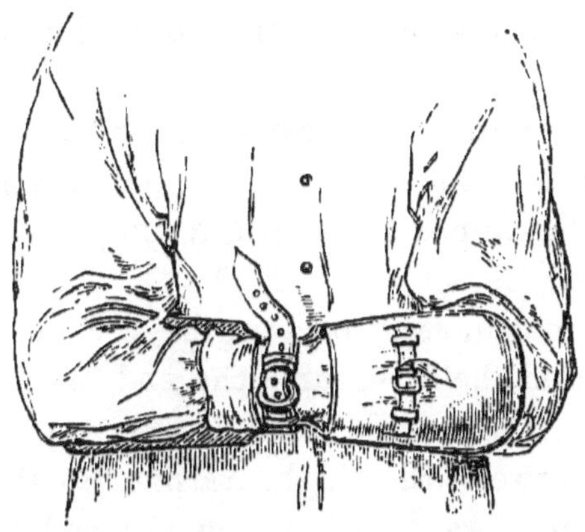

Fig. 10.—Manacles for confining the arms of
delirious patients.

sleeves, and facing the patient, invites him to shake hands. Having thus obtained possession of the patient's hands, he holds them fast while a second assistant, standing behind the patient, pulls the jacket off the first assistant on to the patient, whose hands are thus drawn through the sleeves before he perceives the object of the manœuvre. The jacket is next tied round the neck and behind, the tapes of the sleeves are carried round the body, drawn tight till the arms are folded across the chest, and fastened to the bed on each side, or tied round the body.

Manacles for Delirious Patients.—Instead of the strait jacket a double leathern muff is sometimes used to restrain unruly patients. It irritates them less, and is far more easily applied (see fig. 10).

In wearing it the arms are crossed in front, and a strap is drawn tight round both wrists. Each hand is thrust into a stout leathern glove, or muff, connected with the wrist-strap, and capable of being tightened over the fingers by a strap and buckle across the glove.

To suspend the Testicles.—Suspensories are made specially for this purpose, but a very efficient one can be improvised with a puggree or muslin neckerchief.

Take a muslin scarf or puggree, 3 yards long, wind it once round the hips, carry the end once round the bandage in the middle behind, then bring the scarf forwards between the thighs, raising up the testicles as it comes forward, and attach the end in front to the band round the hips.

When the patient is recumbent, the testes may be supported by a strip of diachylon plaster 2 feet long and 4 inches wide, passed across from hip to hip underneath the scrotum and testes, which lie supported on a shelf.

Another way of raising the testes is to place a soft pincushion between the thighs, and allow the swollen gland to rest on the cushion.

UPPER EXTREMITY.

Bandage for the Finger and Thumb.

Apparatus.—A ¾-inch-wide roller.

The fingers are bandaged to prevent œdema when splints are tightly attached to the fore or upper arm. A roller ¾-inch wide is passed once round the wrist and then carried over the back of the hand to the little finger; then wound in spirals round it to the tip and returned up the finger. The bandage is completed by a figure of 8 round the wrist and the root of the finger, and returned to the wrist. The roller is then brought across the back of the hand to the next finger, to which it is applied in the same manner. The process is repeated until all the four fingers are co-vered. To prevent the bandages from chafing the tender skin, it is a good precaution to place a shred of cotton wool between each finger before carrying the figure of 8 turn round the root.

The thumb is bandaged rather differently: the roller is com-menced in the same way round the wrist, but the first turn is carried at once beyond the last

Fig. 11.—Spica for the Thumb.

joint, turned once or twice round the last phalanx, and continued by reverses to the first knuckle joint; the ball of the thumb is then covered by figures of 8 round the thumb and wrist. This bandage is called the *spica for the thumb,* and is sometimes employed to compress bleeding wounds of the ball of the thumb, when it is applied without previously covering the phalanges, as in fig. 11.

The Hand and Arm.

Apparatus.—1. A roller 2¼ inches wide for an adult but narrower for a child. 2. Some cotton wool.

A little cotton wool should be placed in the palm before applying the roller. The bandage commences with 2 spirals round the wrist, passing at the back of the wrist from the radius to the ulna. The roller is then carried across the back of the hand from the radial border to the root of the little finger (see fig. 12), and then across the palm, reaching the back of the hand between the thumb and forefinger.

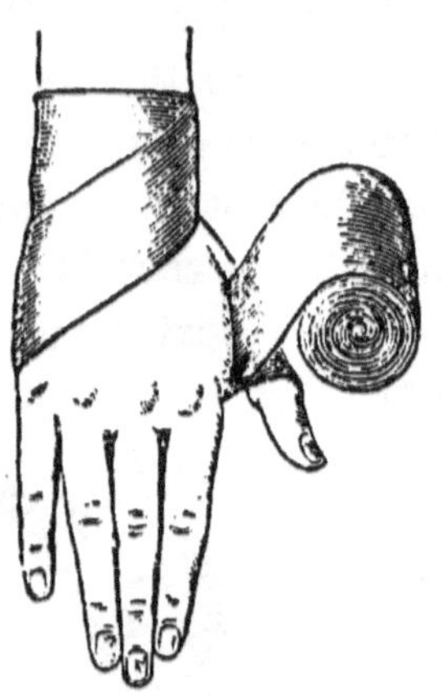

Fig. 12.—Commencing to bandage the hand.

When the hand is covered by these figures of 8 the bandage is passed up the *forearm* by reverses placed over the extensor muscles till the elbow is nearly reached. Before going further a dossil of cotton wool is placed in the bend of the elbow, and on the inner condyle ; the joint is bent to the degree that will be required by the splint, and the patient told to grasp some part of his dress, or the sleeve of the other arm, that he may not unconsciously extend the joint again while the bandage is being rolled round it.

The elbow is covered by first carrying the roller round the joint, so that the point of the olecranon rests on the centre of the turn (see dotted lines fig. 13). The bandage is then continued in figures of 8, passing above and below the first turn until the elbow is covered in and the bandage of the forearm is completed.

The arm is covered by spirals and reverses till the armpit is reached.

Before bandaging *the*

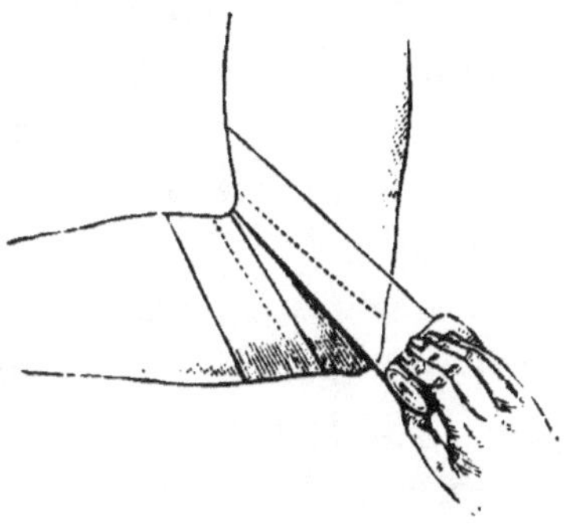

Fig. 13.—Bandage covering the elbow. The first turn over the point of the elbow is shown by the dotted lines.

shoulder the armpit is protected by cotton wool or a double fold of soft blanket ; the roller is then carried in front of and over the shoulder, across the back to the opposite axilla, where also some wool should be placed, then across the chest to the top of the shoulder again, and under the armpit to the front (see fig. 14). These figures of 8 are repeated as often as necessary to complete the covering. The bandage is applied in this method for dressings ; but

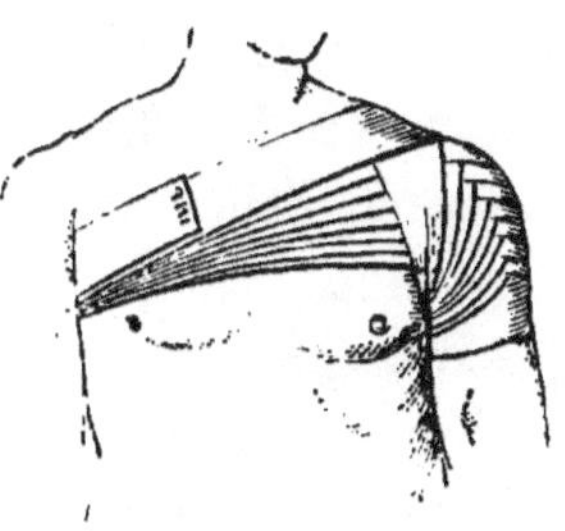

Fig. 14.—Spica Bandage for the shoulder.

when pressure is needed the first turn may be carried at once to the root of the neck, and each succeeding turn made to overlap below the last, until the point of the shoulder is gained.   This bandage is called *the spica for the shoulder*.

Wound of the Palmar arch.—Bleeding from this wound can usually be stopped by pressure on the bleeding point; when this fails an attempt should be made to tie the vessel at the wound, and if this be impracticable the arteries of the forearm must be deligated.

For compression the following apparatus is necessary:—1. Petit's tourniquet. 2. Straight wooden splint. 3. Rollers 2 inches wide, and ¾ inch wide for fingers. 4. Pad and cotton wool. 5. Lint. 6. A slip of a wine cork. 7. Scissors and needle and thread.

Step 1. Apply the tourniquet to the brachial artery, to control the hæmorrhage while the apparatus is being adjusted.

Step. 2. Make a graduated compress by folding a sixpence or disk of a cork in two or three thicknesses of lint, trim the lint into circular disks and prepare a dozen similar disks of increasing size; lay these one on each other to form a round cone about one inch high with the piece of cork at the apex, and fasten them together by a stitch.

Step 3. Clean and dry the wound.

Step 4. Bandage the fingers and thumb, and prepare the splint, which should be straight, as broad as the forearm, and long enough to reach from one inch below the elbow to the tips of the fingers; it should be lightly padded.

Step 5. Envelope the wrist with a little wool; next lay the graduated compress on the wound, the small end downwards, and maintain it firmly in position with the left thumb, while the splint is applied to the back of the hand and forearm. The splint is then fixed by a roller carried in figures of 8 round the hand and wrist across the compress until that is tightly pressed into the wound, when the roller is continued along the forearm. A fold of wool is laid in front of the elbow, the tourniquet removed, and the roller is carried to the armpit while the forearm is raised, flexed across the chest, and fastened to the side.

This apparatus is worn without being disturbed for three or four days if bleeding do not return; but at the end of this time it should be examined; if there be pain, or if discharge ooze out at the wound, the bandage should be removed and readjusted less firmly than before, a piece of wet lint replacing the graduated compress.

Venæsection.—Bleeding and bandage at the *bend of the elbow. Apparatus.*—1. Lancet. 2. Tape. 3. Pledget of lint. 4. Dish. 5. Staff.

In opening a vein at the bend of the elbow, the median basilic is selected, simply because it is usually the largest, but any branch that is superficial, and well filled with blood, may be opened.

The patient should sit or stand, in which positions, faintness, one of the objects of bleeding, is attained by the abstraction of a less amount of blood than in the horizontal posture.

The surgeon places a graduated bleeding dish on a chair or stool within his reach, and a pledget of lint in his waistcoat pocket; he next gives the patient a heavy book or staff to grasp in his hand. The arm being bare to the shoulder, a tape, ¾ inch broad and 1¼ yard long, is tied round the arm tight enough to impede the venous, but not the arterial flow.

The surgeon standing opposite his patient and grasping the arm to be bled with his left hand, so that his thumb controls and steadies the swollen vein below the pro-posed incision, takes his lancet between the right forefinger and thumb ; then going through skin and vein at one stroke, carries the lancet up-wards for about ¼ inch along the vein. The puncture of the lancet should be quite vertical, and the extraction also be made vertically, that the slit in the vein may correspond to the slit in the skin.

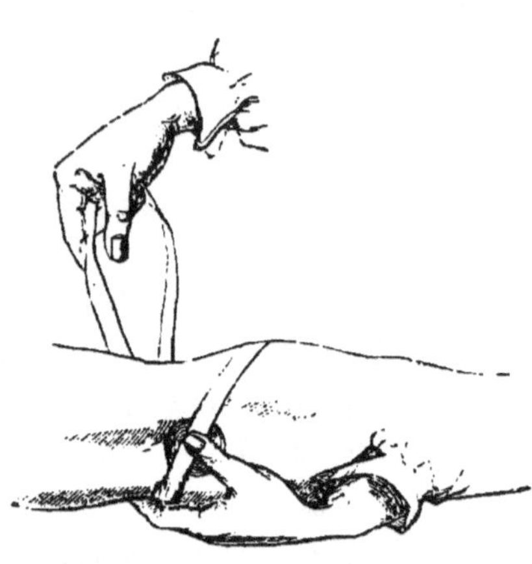

Fig. 15.—Adjusting the tape after bleeding.

This being done, the operator lays aside his lancet, and takes up the dish, holding it so that the blood will flow into it : when the dish is placed, he lifts his left thumb from the vein cautiously or the sudden spirt of blood will fall outside the dish and be lost. When the desired amount is drawn, the operator compresses the vein again with the left thumb, and setting down the dish, puts the pledget of lint over the wound. He keeps the pledget in place with his

left thumb, while he releases the tape round the arm
and places its middle obliquely across the pledget.
His left thumb presses the pledget on the wound,
while the right hand takes the end of the tape which
is farthest from his left, and passes it under the forearm below the elbow to his left fingers, which grasp it
tightly. He then takes the other end with his right
hand (see fig. 15), and bringing it round the arm
above the elbow, carries it across the pledget : as he
does this, he replaces his left thumb on the compress
by his right forefinger, which he keeps there while he
brings up the end of the tape he has already in his
left fingers, and throwing it over the arm *above* his
right forefinger, lets it go ; then passing his left hand

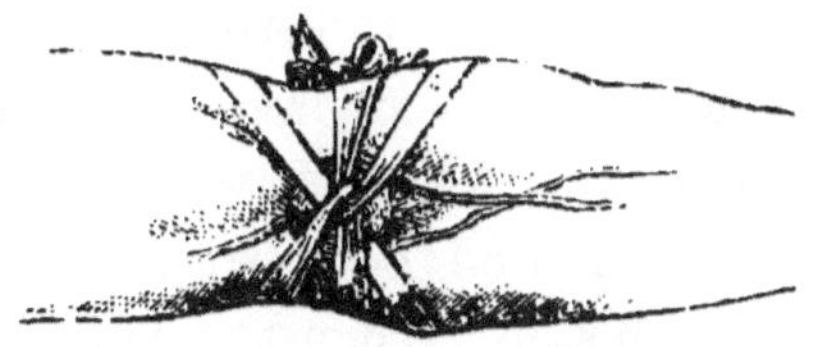

Fig. 16.—The bandage completed.

below the right forefinger, he catches the same end of
the tape again and draws it back. The two ends thus
locked in a loop over the compress, are secured by
tying them in a bow outside the elbow and the operation is finished (see fig. 16).

THE LOWER EXTREMITY.

For adults the most useful width for the rollers is
3 inches, and the length the ordinary one of 8 yards.

The Foot is usually bandaged without covering the heel, and the bandage is begun as follows :—

The roller being held in the right hand for the right foot, or in the left hand for the left foot, the unoccupied hand takes the end, and lays it across the inside of the ankle. The roller is then brought forwards to the root of the little toe where it passes under the sole to the inner side of the foot just behind the great toe. By carrying the roller across the foot again to the outer side, the loose end is made fast. When two turns have been carried round the foot, the bandaging is continued by reverses until the fore part of the foot is covered ; then one or two figures of 8 round the foot and ankle carry the bandage to the *leg*, where it proceeds upwards by spiral turns round the small of the leg, and by reverses round the calf. The reverses lie at equal distance up the outside of the leg, on the muscles, not over the shin, that the skin be not pressed between the crease of the bandage and the bone. When the calf is passed, the roller is carried by figures of 8 above and below the knee, until that joint is covered in, then continued by reverses up the thigh to the groin, where the bandage terminates by a spica round the body (see p. 10). This is the ordinary bandage for the lower limb. There are some varieties for particular parts ; these are :—

To cover the Heel.—Holding the roller as for the foot, carry the end behind the heel outwards to the front of the ankle, and bring the roller by the inside over the front of the ankle-joint, to complete the turn. In doing this, the point of the heel must catch the middle of the band. If the foot is a long one, the roller should be three inches broad; but a narrower bandage is more easily fitted on to a small foot. After the first turn, the roller is carried in figures of 8 round the foot and ankle, passing alternately, as in fig. 17, above and below the first turn until the joint is covered.

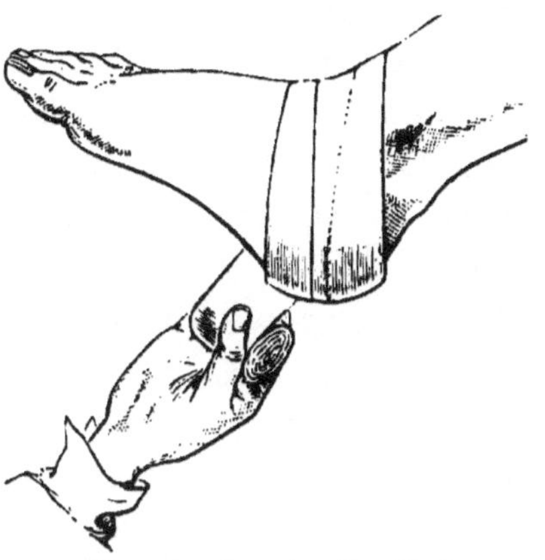

Fig. 17.—Covering the Heel.

To bandage a Toe. — Take two turns round the foot, with a bandage one inch wide, then pass it round the toe to be raised, and back again round the foot. This figure of 8 lifts a toe above the rest if taken from the dorsal, and depresses it if taken from the plantar surface.

The Knee is bandaged by beginning with a simple turn round the leg above the calf, then carrying the roller across the patella to the thigh above the knee, where a circular turn is taken round the thigh before descending over the patella to the leg below the knee; this process is repeated until the knee is covered.

To bandage a Stump.—The flaps are first supported by two or more strips of plaster, one inch wide and ten or twelve inches long, carried from the under surface of the limb over the face of the stump, and a slip of wet lint and oiled silk is applied to the wound. The muscles and soft parts are next confined by a bandage.　This is first fixed by simple turns below the nearest joint, and brought downwards in figures of 8 round the limb till the end of the stump is reached, which is next covered in by oblique and circular turns carried alternately over the face of the stump and round the limb, as is shown in fig. 4 for bandaging the head.　If a double-headed roller be used, that is applied in the manner directed for the capelline bandage on p. 6.

Extending a Stump.—When the soft parts fall away from the bone, they may be drawn down by attaching a weight by cord and pulley, as described for extending the hip-joint (see page 86).　The stump should be lightly bandaged and the cord be connected with each flap by an extension stirrup (see p. 78).　The weight is at first one or two pounds, and should be increased from time to time as required.

Many-tail, or 18-tail Bandage, or *bandage of Scultetus.*—A roller is cut into short lengths long enough to encircle the limb and to allow the ends to overlap 3 inches; they are applied separately, so that each successive strip covers half of its predecessor.　Sometimes the tails are attached to each other before they are used.　When this is done, they are so laid out on a table that the second overlaps one half of the width of the first, while the third overlaps the second to the same extent, and so on.　A vertical strip is next laid across their middle, and fastened to each tail by a stitch; but this arrangement is not a necessary part of the bandage, and it prevents single tails from being removed.　This bandage is used in compound fractures and other wounds, as the soiled strips can be replaced without raising the limb to pass the roller under it.

Elastic Socks and Stockings are made to support varicose veins of the legs. They are woven of silk or cotton webbing with india-rubber threads. The cotton ones are lower in price and often even the more comfortable to wear. The stockings should fit carefully, especially at the small of the leg, where they generally are too slack, while they clip tightly at the upper end below the knee.

CHAPTER II.

STRAPPING.

STRAPPING is a method of supporting weak or swollen joints and other parts. Sheets of calico, holland, wash-leather, or white-buckskin, spread with lead or soap plaster, are used for this purpose.. The American india-rubber and isinglass plasters are for certain purposes valuable improvements.

Before a sheet of plaster is used, it should be rubbed with a dry cloth, to remove adherent dust, &c. It is then cut into strips varying in width between $\frac{3}{4}$ inch and 2 inches, according to the evenness of the surface to be covered : narrow strips fit best over joints and irregular surfaces. When applied to a limb, the strips should be about one-third longer than its circumference. Each strip or strap is first warmed by holding it to a fire, or by applying its *unplastered* side to a can of boiling water; when hot, the strip is drawn tightly and evenly over the part. If the surface to be strapped be irregular, it is best to dip each strip of plaster into hot water before applying it ; being thus rendered quite supple the strap fits the limb more exactly. When the limb is thickly beset with hairs, it is a good plan to shave the part where the plaster will lie before putting on the straps.

Strapping the Breast.—Strapping is put on the breast in the same way as the bandage (page 9). The straps should be not more than 2 inches wide, and long enough to pass forward under the axilla and breast from the lower angle of the scapula on the side of the injured breast, and then mount across the chest to the shoulder as far as the spine of the other scapula. The strips are warmed and laid on alternately over the breast and across the chest, until the former is fairly supported.

Strapping has this advantage over a bandage—its circular strips do not pass completely round the chest and thereby hamper the breathing as the roller does.

To Strap the Testicle.

Apparatus.—1. Strips of soap plaster spread on calico, or better, on wash-leather, $\frac{1}{2}$ inch wide and 12 inches long. 2. A can of boiling water. 3. Razor and soap.

First, shave the scrotum ; then tighten the skin over the testis with the left thumb and forefinger passed above it ; take a strip of plaster 6 inches long and $\frac{1}{2}$ inch wide, and encircle the cord tightly with

Fig. 18.—Strapping the Testis.

it ; next pass another strap of the same width, 9 or 10 inches long, from the back of this ring, over the testicle to the front, drawing it tight also (see fig. 18).

The strapping is continued by applying fresh straps which overlap each other until the whole testis is covered in. Lastly, take a strip 15 or 20 inches long, and, beginning at the ring above, wind it round and round the testicle until all the vertical strips are confined in place by this spiral one.

The strapping should be re-applied the second or third day, as the testicle by that time will have shrunk within its case.

Strapping Ulcers and Joints.—Cut strips of plaster one-third longer than the circumference of the

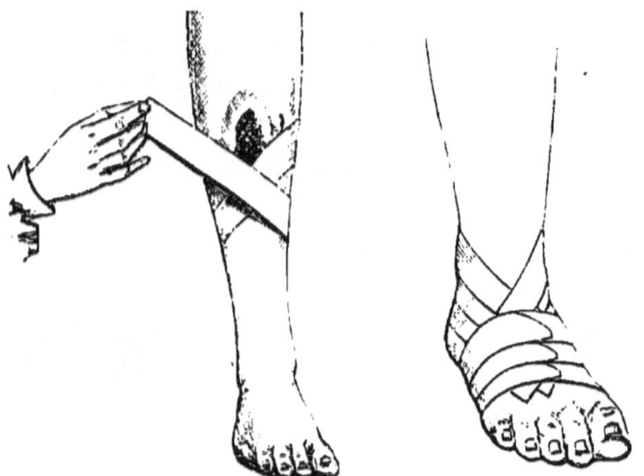

Fig. 19.—Strapping an Ulcer.   Fig. 20.—Strapping the Ankle.

part to be strapped ; if that is irregular, as the ankle or wrist, they must be narrow : commonly the width varies between $\frac{3}{4}$ inch and $1\frac{1}{2}$ inch. The strips are warmed, the middle passed behind the limb, the ends crossed in front (see fig. 19), and drawn tight, but with sufficient obliquity for the margins of the strip to lie evenly. The strapping is begun as low down the limb as requisite, and continued upwards by laying on more strips, each overlapping about two-thirds of the preceding strap. When the process is finished, the ends should meet along the same line, and all the uppermost ones be on the same side.

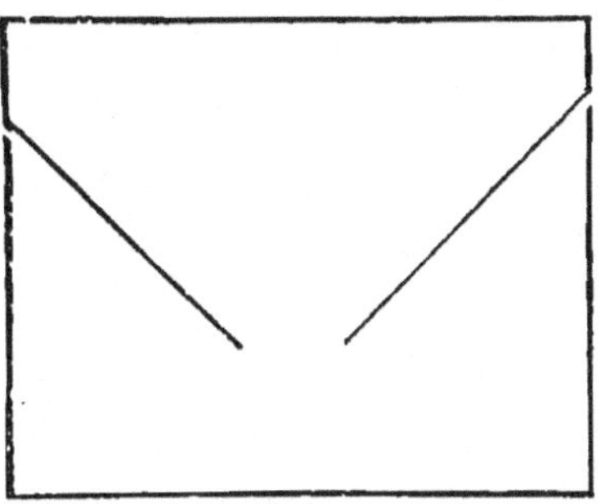

Fig. 21.—Plaster for Ankle.

The ankle is strapped differently (see fig. 20). Strips are prepared about 1 inch wide; one is carried behind the heel and its ends brought forward till they meet on the dorsum of the foot; a second, encircling the foot at the toes, secures the first; a third is again carried behind the heel above the first, and is fixed by a fourth round the foot. This is continued until the foot and ankle are firmly supported.

To strap the Ankle with one piece of plaster.—The following

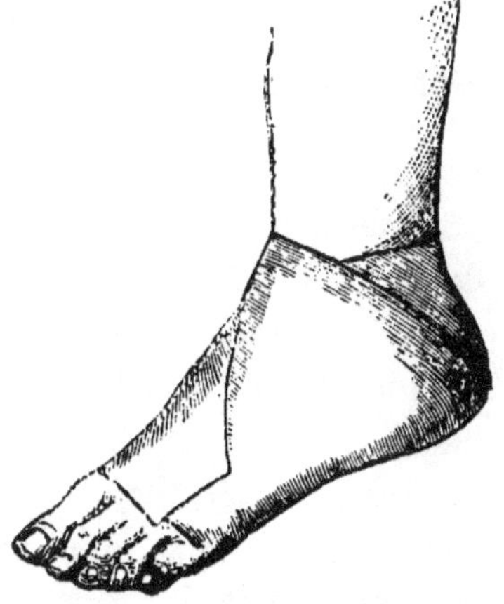

Fig. 22.—Ankle plastered.

measurements are required :—1. The first begins at the back of the leg just above the malleoli, and passes behind the heel along the middle of the sole to the head of the 5th metatarsal bone. 2. The second is carried round the ankle-joint at the point of the heel. Next, cut a piece of diachylon plaster spread on stout linen or buckskin leather of the required dimensions (usually about 10 inches by 13 inches in the adult male), and cut from the shorter sides a slit running diagonally from a point 2 inches from the base to the point where the heel will rest (see fig. 21). The foot is put on the plaster, and the two upper flaps are brought forward round the ankle, care being taken to wrap well round the joint, making the plaster fit smoothly over the uneven surface; the lower flaps are afterwards brought up on each side to the dorsum from the sole, and complete the covering as shown in fig. 22.

Strapping a Joint with Mercurial Ointment.
(*Scott's Bandage.*)

Apparatus.—1. Mercurial ointment. 2. Diachylon plaster. 3. Lint. 4. Spirit of camphor. 5. Cotton wool. 6. Freshly scalded starch, or solution of gum. 7. Binder's millboard. 8. Rollers two or three inches wide.

Spread the ointment on a piece of lint large enough to envelop the joint, and to extend four or six inches above and below it; then wash the joint with warm water and soap, and dry it carefully; next sponge it well with the spirit of camphor for five minutes. Tear the lint into strips and wrap them round the joint; then strap the part firmly from below upwards over the lint with strips of diachylon plaster, each over-

lapping the preceding one.   Lastly, envelop the joint
in a thin layer of cotton wool, and roll a bandage
soaked in starch over all.   If the patient wears no
other kind of splint the bandage may be strengthened
by laying a piece of millboard well softened in boiling
water along each side of the joint before the starch
bandage is applied.   As the enlargement of the joint
shrinks, this application must be renewed; usually
every fortnight is often enough.

CHAPTER III.

—◆—

FRACTURES.

HEAD AND TRUNK.

Fracture of the lower Jaw. — The External Splint and Bandage. — A method requiring the lower jaw to be supported against the upper one while the broken bone knits.

Apparatus.—1. One and a half yards of bandage four inches wide. 2. A piece of gutta-percha, sole-leather, binder's millboard, or poroplastic felt. 3. Dentists' silk or wire. 4. Boiling hot, and cold water. 5. A punch. 6. A pad or fold of blanket.

Step 1. The fracture is first reduced. While the apparatus is being fitted, the recurrence of the displacement is prevented by the hands of an assistant, or by lacing the teeth together with stout silk or wire. It is well also to wet the patient's chin with a sponge and cold water, to prevent the gutta-percha from sticking to his beard while it is soft.

Step 2. A piece of gutta-percha is prepared $2\frac{1}{2}$ inches wide and long enough to reach from one angle of the jaw to the other when passing in front of the chin. This is softened thoroughly by immersion in *boiling* water, and when quite pliable should be quickly removed from the hot and plunged for a moment into cold, water; the operator's hands should also be wetted. While softening the sheet, care must be taken not to stretch it. If a towel be previously laid in the hot basin, the gutta-percha can be lifted out without stretching, and laid on a table, while the surface is sponged with cold water to prevent its sticking to the skin; it is then slit from each end into tails 1 inch and 1½ inch wide, leaving 2 inches unslit at the centre. So prepared, the splint is applied to the jaw with the middle pressing against the chin, the narrower ends being carried horizontally backwards to the angles of the jaw; the broader part is next bent up beneath the chin, its ends overlapping the horizontal ones. While the splint is still soft, the surgeon presses it firmly upwards that the gutta-percha

may mould itself accurately to the chin. When set, the splint is removed, trimmed, cut away a little opposite the throat, and punched with holes here and there for perspiration. A covering of wash-leather may be added, if desired. When the splint is finished, it is replaced on the chin. If sole leather or pasteboard be used instead of gutta-percha, they must

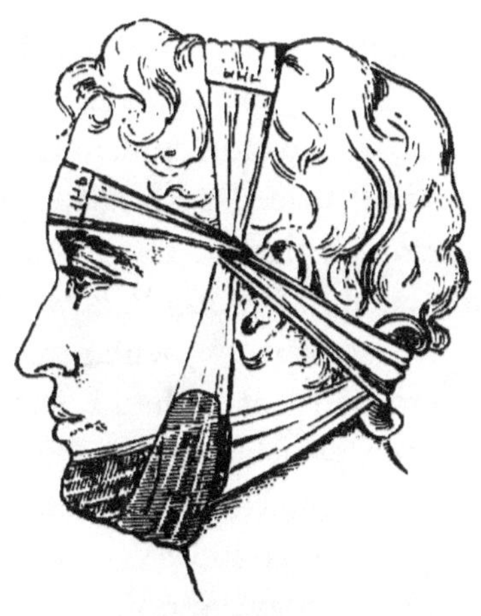

Fig. 23.—Outside splint for fracture of the lower jaw.

be prepared in the same way, but allowed to remain on the chin twenty-four hours that they may set before the final trimming and adjustment.

D

Step 3. A bandage, 4 inches wide and 1½ yards long, and slit from each end up to about 2 inches from the centre, is then applied to the splint, and a small pad of folded flannel should be placed at the nape of the neck to protect the skin from the crossed bandage. When all is ready, the two upper ends are carried behind the neck, drawn tight, and tied in a knot ; the lower ends are taken upwards, and fastened at the vertex (see fig. 23).

The loose ends of the bandage already tied at the nape, are brought forward, and passing the upright bandage at the temples are pinned or stitched firmly together on the brows, and to the upright band at the temples.

The ligatures that may have been used on the teeth can now be removed, or if they cause no pain, they may be left for a week.

It is a useful precaution to place a piece of soft leather spread with soap plaster under the chin and along the throat, to protect the skin from the chafing of the splint while it is worn.

On emergency, when gutta-percha, leather, or pasteboard are not at hand, the jaw may be set, and then kept in position by a four-tail bandage, made from a pocket-handkerchief, until more complicated apparatus can be prepared.

The apparatus must be worn five weeks before it is laid aside and mastication permitted.

Interdental Splints.—In cases of unusual difficulty, interdental splints may be employed. But to fit them properly, the mechanical skill of a dentist is requisite.

A fractured Rib is very well treated by strapping the injured side alone, without enrolling the chest in a tight bandage, which harasses the patient by impeding respiration if the arm is fixed to the side.

Apparatus.—1. Diachylon plaster. 2. Can of boiling water. Strips of plaster long enough to reach from the spinal column to the sternum, and 2 inches wide, are to be firmly drawn round the injured side. The first strip should be

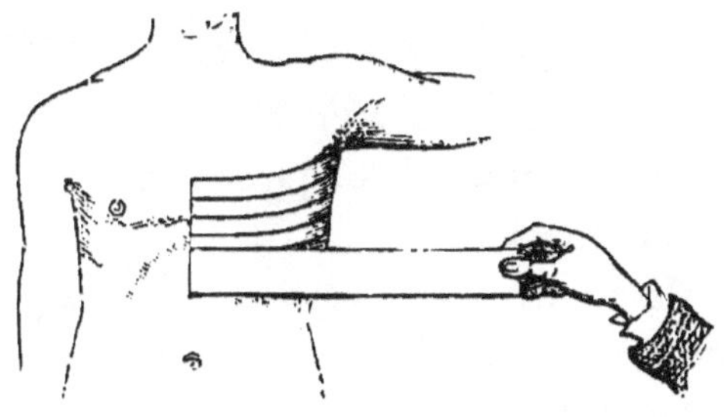

Fig. 24.—Strapping a broken Rib.

carried as high as can be managed under the armpit. The next strip overlaps it about an inch (fig. 24), each succeeding strip overlapping and fixing the preceding one until the lower ribs are covered in. The arm should then be bandaged to the side, and supported in a sling.

A *second mode* of treating fractured ribs, is to take a flannel roller 6 inches wide, and 8 yards long, and carry it firmly round the chest in successive spirals, beginning at the armpits, and passing down till the waist is reached. The turns of the roller may be kept from slipping down by throwing across the shoulders two strips of bandage like a pair of braces, and stitching each turn to the braces in front and behind. The arm should be confined to the side as in the other method. This plan has the inconvenience before mentioned of interfering with respiration.

In Fracture of the Pelvis, the fragments are kept in position by a broad roller carried several times round the pelvis and fastened. Or, more effectually, by encasing the hips with starch bandage, or plaster of Paris ; the case should be continued down both thighs one-fourth of their length.

THE UPPER EXTREMITY.

Fracture of the Metacarpal Bones.

Apparatus.—1. A piece of gutta-percha. 2. A roller 2 inches wide.

In treating this fracture it is important to keep the broken bone in place without confining the wrist or fingers.

A pattern of the palm and dorsum of the hand is cut out of paper, which is doubled round the radial side, letting the thumb out through a hole of convenient size to clear it (see fig. 25). The piece of paper is then laid on a sheet of gutta-percha $\frac{1}{4}$ inch thick, and the requisite quantity cut off. A hole as big as

Fig. 25.—Gutta-percha Glove for fractured Metacarpal Bone.

a pea is next punched in the gutta-percha in the middle, about 1 inch from the lower border, or at a point corresponding to the hole in the paper for the thumb. The fragments are then pushed into place and held so by an assistant, while the surgeon softens the gutta-percha in boiling water (see p. 49) ; when thoroughly soft, he draws the thumb through the little hole punched in the gutta-percha, and moulds the splint to the palm and back of the hand, bringing the ends of the gutta-percha together at the ulnar side of the hand ; the fragments are held carefully in position till the splint is set. It is important that the gutta-percha be pressed well into the hollow of the palm, in order that the broken ends of the bone may be kept in their places. The splint is afterwards removed and trimmed. A few holes should be punched in it after it is moulded to allow perspiration to escape. The splint may then be covered with wash-leather, and a pair of straps with buckles stitched on to keep it in place. It is worn for three or four weeks, or until the fragments are united.

Another plan is effectual should gutta-percha not be at hand.

Apparatus.—1. A firm ball of tow large enough to fill the palm, stitched in old linen. 2. A roller 2 inches wide.

The broken bone is first replaced; then the hand and fingers are bound on to the ball by carrying the roller around them until they are all immovably confined.

This plan has the disadvantage of confining the whole hand for the fracture of one metacarpal bone ; the gutta-percha allows the use of all bones but the metacarpals.

Broken Phalanges are treated by bandaging them on to a slip of wood long enough to reach into the palm ; the slip must be well padded, that the somewhat concave anterior surface of the digit may be accommodated to the flat splint.

If more than one finger be injured, and the fracture

be compound, the splint should reach up the palmar aspect of the hand and forearm. Fingers should be cut in the splint to correspond with the injured fingers.

A trough or stall of gutta-percha moulded to the broken finger is better than a wooden splint for keeping the fragments in position.

Fracture of the lower end of the Radius.— *Colles' Fracture.*—The displacement in this fracture is mainly due to the lower end of the radius and the carpus being carried backwards while the shaft projects in front.

Fixed deformity opposite the wrist is usually present from impaction of the fragments ; moderate extension may be employed to remove this, but forcible or continued efforts give great pain and do harm, by straining the already torn ligaments still further.

The objects to be attained in treating this fracture are to press the lower fragment forwards while drawing (adducting) the hand towards the ulnar side of the limb. For this purpose a straight and a curved splint are used.

No bandage should be placed under the splints in treating any fracture of the shafts of the radius or ulna, lest the broken ends be pressed into the interosseous space.

Apparatus.—1. A straight splint of wood. A second splint, curved at its lower end. 2. Pads and cotton wool. 3. A roller 2 inches wide. 4. A sling. 5. A strip of plaster. 6. Pins.

Step 1. Prepare the splints. The straight splint should reach, when the elbow is bent to a right angle and the thumb is upwards, from a little below the inner condyle to the lower end of the upper fragment or shaft ; the curved or pistol splint extends from the outer condyle to the joint of the first and second phalanges. The width of both splints should slightly exceed that of the forearm. The bend of the lower end of the pistol splint should be abrupt, and directed

towards the ulnar border opposite the wrist, where the margin of the splint should make an obtuse angle of about $1\frac{1}{2}$ right angles (see fig. 26).

Fig. 26.—Pistol Splint for fracture of the Radius near the lower end.

Pads used with these and other wooden splints are made of layers of cotton wool, carded sheep's wool, tow, or folds of old blanket. These materials should be stitched in old linen or calico, and covered outside with oiled silk where likely to be stained with the discharge from wounds.

The pads must be thicker below than above, to keep the splints parallel along the forearm : and that of the pistol splint is thickest opposite the carpus, to push the lower fragment forwards.

After these preparations the splints are applied.

Step 2. Put a very little cotton wool in the palm and across the root of the thumb, before the roller is begun, lest it chafe the carpus in front. The curved splint, with the barrel or longer part inclined downwards below the forearm, is next attached to the back of the hand by a roller carried in figures of 8 round the hand and root of the thumb, but not above the wrist (see fig. 27). This is made fast by a pin.

Step 3. Raise the straight part of the outside splint till parallel to the forearm, thus adducting the hand to the ulnar side ; and fix the splint by a strap of plaster an inch wide carried round it and the forearm below the elbow.

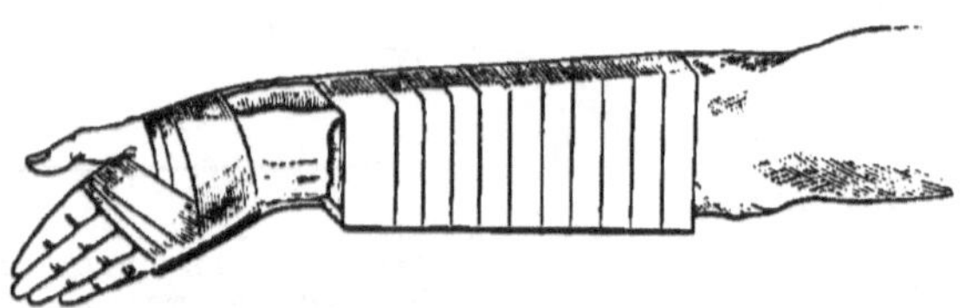

Fig. 27.—Fracture of the Radius.

Step 4. Apply the inside straight splint next, keeping the front of the carpus and of the lower fragment exposed. Draw the two splints together by simple spiral turns of a roller, begun just below the elbow and carried down to the lower end of the inside splint, there fasten it off.

Step 5. Put a narrow sling under the forearm between the elbow and the wrist to support the limb comfortably.

When the apparatus is finished the position of the broken fragments should be visible (see fig. 27) and not concealed by bandage. The hand should also be quite free of the sling, lest it be drawn from its proper position of adduction. The fragments are in good position when the hollow on the anterior aspect of the wrist and the prominence on the corresponding posterior surface are removed.

The Gutta-percha Gauntlet is another plan of treating fracture of the lower end of the radius that

may often be adopted from the first, and may always
replace the wooden
splints and band-
age when the
swelling has sub-
sided. The gaunt-
let is moulded to
the hand and fore-
arm (see fig. 28)

Fig. 28.—Gutta-percha Gauntlet for
Colles' Fracture.

in a manner similar to that described for making a
sheath for fractured metacarpal bones.

Carr's Splints possess several advantages over the
pistol splint. The splints are made in right and left
pairs of stiff oak or beech, of the following dimensions
(see fig. 29). Along the surface of a strip of hard
wood, $\frac{1}{8}$ inch thick, $1\frac{3}{4}$ inches wide and 11 inches long,
is fixed a piece of wood 9 inches by $1\frac{1}{2}$ inches, shaped
to fit the hollowed anterior surface of the radius; being
$\frac{1}{2}$ inch thick at the radial side and bevelled to a feather

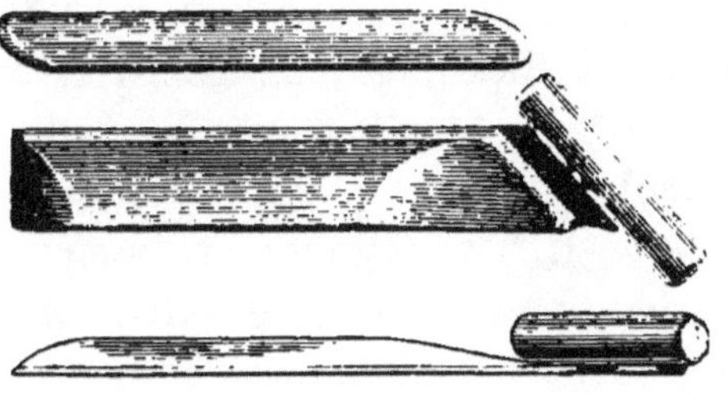

Fig 29.—Carr's Splints.

edge at the ulnar side. Across the lower end of the
splint a cylindrical cross-piece $\frac{3}{4}$ inch thick and 4 inches
long is screwed obliquely towards the ulnar side. A
second strip of wood. 9 inches long, $1\frac{1}{2}$ inches broad

and $\frac{3}{16}$ inch thick, completes the pair of splints. Each strip is lightly padded by a threefold thickness of lint.

When applying the splints, the larger one is laid along the front of the forearm so that the cross piece lies in the palm exactly opposite the knuckles. The fracture is reduced, and the patient made to grasp tightly the cross-bar. The plain splint is then applied from the knuckles upwards, and the two bandaged together. The fingers and thumb are left free, while the wrist is firmly fixed between the two splints. The patient should be encouraged to use the fingers as freely as possible while the bones are knitting.

By the two last plans the fingers are left free, and some motion allowed also to the thumb. The only joints kept immovable are those of the carpus and wrist.

Apparatus of some kind must be worn for three weeks continuously; then for a fortnight longer interruptedly, being removed every day to allow passive motion to be applied to the fingers, gradually extending to the wrist also. The patient should be warned that pain and stiffness last long after this fracture.

Fracture of the Shaft of one or both Bones of the Forearm.

Apparatus.—1. Two straight wooden splints. 2. Pads and wool. 3. 2-inch wide roller. 4. Sling.

The treatment is the same whether one or both bones are broken. Caution has been already given against bandaging the forearm underneath the splints.

Step 1. Prepare two straight wooden splints; one to go in front of, and one behind the forearm. The posterior or outside splint reaches from the external condyle to the end of the metacarpus; the anterior

or inside splint from a little below the internal condyle only as far as the wrist, keeping clear of the ball
of the thumb. The splints should be slightly broader
than the forearm, and well padded ; towards the lower
end the padding should be thicker than above. The
forearm is bent to a right angle and the thumb put
upwards.

Step 2. Reduce the fracture by gentle slow extension
at the wrist ; this being effected, apply the splints to
the forearm, and let an assistant hold them while the
bandage is rolled on. Gentle extension of the hand
and wrist should be maintained throughout the application of the roller.

Step 3. When a little wool has been wrapped round

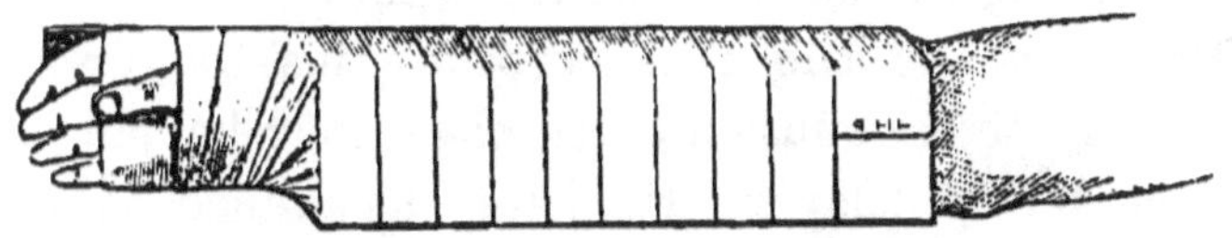

Fig. 30.—Fracture of both Bones of the Forearm.

the hand and wrist, fasten the dorsal splint by figures
of 8 carried round those parts; then draw the two
splints together by simple spirals continued to the
elbow (see fig. 30).

Step 4. Support the forearm in a sling, to complete
the apparatus.

The splints are worn three weeks ; after this, passive
motion may be practised daily, and the splints finally
abandoned ten days later. But a sling is still required
some ten days after the splints are laid aside.

Sometimes the longer splint is laid on the palmar
aspect of the forearm, from 1 inch below the inner

condyle to the tips of the fingers, and the shorter splint from the outer condyle to the metacarpus. If this plan is adopted the padding of the splint opposite the wrist and palm must be nicely adjusted to support the limb and steady it thoroughly.

When the ulna alone is broken, an anterior splint reaching from the inner condyle to the tips of the fingers often suffices without a second one.

When the shaft of the radius is broken high up (a rare accident) the displacement is sometimes very difficult of reduction unless the wrist be well supinated. To preserve this position it may be necessary to use a wooden angular splint, and to fix the vertical part to the arm behind the elbow, while the horizontal part is carried along the back of the forearm.

Fracture of the Olecranon.—This fracture, if seen early before effusion takes place, may be put up at once, but if delay till the joint has swelled occurs, the limb must be kept quiet, with evaporating lotions, on a pillow, or on a splint in an easy position until the effusion is absorbed, before any means can be taken to restore the position of the olecranon. Though the straight position of the elbow is usually employed, it is not essential for even very close union of the fragments.

In treating this fracture the following plan is useful.

Apparatus.—1. Straight hollow splint. 2. Two-inch rollers and finger rollers. 3. Pad, wool, and lint. 4. Strapping plaster. 5. Pins.

Step 1. Bandage the fingers; wrap the hand in cotton wool and bandage it. When the wrist is passed, fasten the bandage for a time by a pin, and straighten the arm.

Step 2. Push the olecranon down as close as possible to the rest of the ulna, and put a dossil of lint above it. Place the middle of a strap of plaster an inch wide and 16 inches long, on the lint, and carry its ends round the forearm in a figure of 8; to some extent this fixes the fragment.

Step 3. Continue the bandage up the forearm by reverses, keeping the elbow straight; wrap some cotton wool thickly round the injured joint, and then cover it in by figures of 8 carried over the compress of lint and the forearm, to draw down the olecranon (see fig. 31). When this is secured, prolong the bandage to the deltoid, to confine the action of the triceps muscle.

Step 4. Pad lightly a hollow splint about 2 inches wide, reaching from the axilla nearly to the wrist,

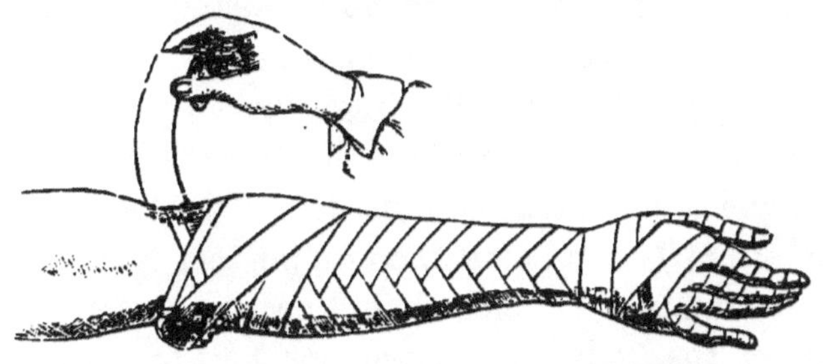

Fig. 31.—Bringing down the Olecranon with Figures of 8.

and apply it along the anterior aspect of the limb, then fix it by a second roller. This completes the apparatus.

The splint and rollers should be removed on the fourth or fifth day, that the position of the fragments may be examined and the roller again applied to draw them closer together.

After twelve days, passive motion of the wrist and fingers, with pronation and supination of the radius, should be adopted, but great care is to be taken that the patient does not inadvertently bend the elbow joint while free of the splint. The splint must be worn, with the frequent removals directed above, for five weeks, during which time gentle flexion of the elbow should be practised after twenty days of complete rest. During passive motion, the surgeon's finger pushes the olecranon against the ulna.

Another Plan.—The method just described is often irksome to the patient. By moulding a sheath of gutta-percha to the arm and forearm, while his limb is straight and the fragment of bone held firmly against the ulna by the surgeon clipping the arm just above the elbow with the thumb and forefinger of his left hand until the soft gutta-percha has set, a fitted splint is formed which retains the fragments in position without tightly constricting the limb; and the bandage for the hand and fingers is not needed. The sheath should reach from 1 inch below the arm-pit to 1 inch short of the wrist, and its edges should nearly meet along the front of the limb. For directions how to trim, perforate and line the splint, the reader is referred to the description of the metacarpal glove, on p. 37.

Fractures of the Humerus near the elbow.— These resemble dislocations of the ulna and radius backwards, but are distinguished from them by the ease with which the bones slip into place and again slip back from it when left to themselves; by crepitus; and, when the fracture is above the condyles,— the common accident,—by those projections retaining their natural relation to the olecranon. In children and youths the articulating surface of the humerus may separate from the shaft without carrying the rest of the lower epiphysis with it. In this rare accident the main distinctions are : from the usual fracture, the projection of the olecranon behind the condyles ; from dislocation, the absence of the hollow of the sigmoid notch, and facility of reduction.

In ordinary cases, where the deformity is reduced without much difficulty, and the injury to the joint is not severe, lateral rectangular splints of leather, hollowed wood, or wire gauze, answer very well. These are placed both inside and outside the limb, and reach from the axilla and shoulder to the wrist. They are applied in the following manner :—

Apparatus.—1. Lateral hollowed angular splints. 2. Pads and wool. 3. Rollers 2 inches wide for the arm and 1 inch wide for the fingers. 4. Sling.

Step 1. The splints must be prepared.

Wooden and wire gauze splints are double. One, inside the arm, reaches from the axilla to the wrist, the elbow being bent to a right angle. The other extends, on the outside, from the deltoid to the wrist. They are better if provided with hinges opposite the elbow, so that their angle can be altered, if desired, in the latter stage of the treatment. Splints of wood or wire gauze must be evenly and lightly padded before application.

Step 2. Bend the arm to a right angle with the thumb upwards. An assistant next reduces the fracture, and holds the bones in position. Then apply the splints. When adjusting the inside splint, care must be taken that the internal condyle is eased from pressure by sufficient padding above and below it. Next fasten on the splints by a roller begun at their lower ends, leaving the hand free, and carried up to the elbow. Before turning round that joint a soft pad must be placed at the bend of the elbow to push the lower end of the humerus back, and the length of the arm should be measured against the unbroken one to make sure that the shortening is reduced. Extension is kept up the whole time the splint is being fixed to the arm. That is done by carrying the roller round the elbow with figures of 8 and by simple spirals up to the axilla where it is finished off.

Step 3. Lastly, the forearm is supported in a sling under the wrist, leaving the elbow free (as in fig. 34, page 52).

After three weeks of complete immobility, passive motion should be applied to the elbow daily, during the fortnight or three weeks more that the splint is still worn.

If the displacement returns very easily, it is better to use an **L**-shaped splint passing behind the arm and below the forearm. This may be made of wood, or of leather, or of gutta-percha, in the mode about to be described.

The **L-shaped splint** of gutta-percha, or leather, is made as follows :—

Apparatus.—1. Sheet gutta-percha ¼ inch thick. 2. A tray or wide wash-hand basin. 3. A basin of cold water. 4. A kettle of boiling water. 5. A towel. 6. A knife.

Cut from the sheet of gutta-percha a piece long

enough to reach, while the elbow is bent and the thumb turned upwards, from the armpit down the back of the arm and under the elbow and forearm to the wrist. The gutta-percha must also be wide enough for its edges to meet along the front of the limb. Prepare the tray with the hot water, lay in it the gutta-percha, and pour on more, almost boiling, water, adding still more water as the first cools; this may be done by an assistant, while the surgeon grasps the forearm and reduces the frac- ture. When the gutta - percha is soft, the assistant lifts it from the tray, and waits for a moment before laying it on the

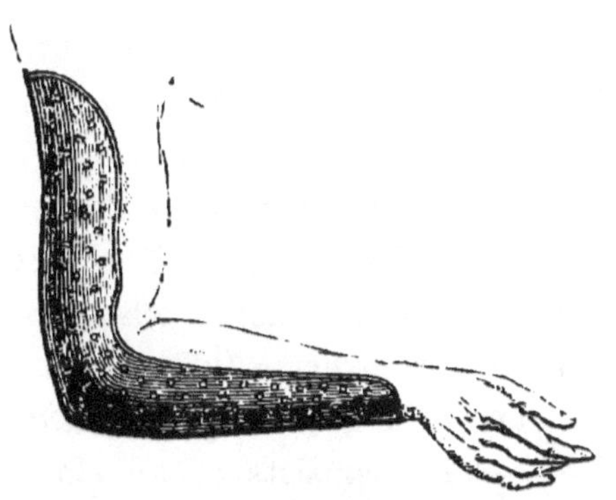

Fig. 32.—Gutta-percha Splint for Fracture at the lower end of the Humerus.

limb. While the gutta-percha is setting, the elbow is kept at a right angle, with the broken bone in place, and the edges of the soft gutta-percha should be pinched together along the front of the arm to obtain an accurate mould of the part. When it is set, the splint is removed to be trimmed, perforated, and covered with wash-leather. It is then ready for use.

Leather takes so much time to set that it should not be used in recent fractures. When the bone is partly set, leather is a useful substitute for wood.

The best way to fit leather is to take a plaster cast of the limb from a gutta-percha mould made in the way just described, and to bandage the wet leather to the cast until it has dried. It is first cut from a sheet of sole-leather in the same manner as the gutta-percha, but is trimmed before soaking, not after it has been moulded, like gutta-percha. Moreover, leather must be notched at the elbow to enable it to conform to the shape of the limb. If possible it should have twenty-four hours' soaking in *water* before being fitted to the cast or to the limb itself ; but when this cannot be done, immersion in hot water, into which a teacupful of vinegar has been thrown, will make the leather quite supple in a quarter of an hour. The leather splint must be worn twenty-four hours while it sets, and then be removed for covering (see Leather Splints).

Stromeyer's cushion for gunshot injuries about the elbow.—This consists of an elastic cushion of horsehair, triangular in shape, and thicker at the base than at the apex (see fig. 33). When in use the narrow apex is placed in the axilla, and then fixed by two

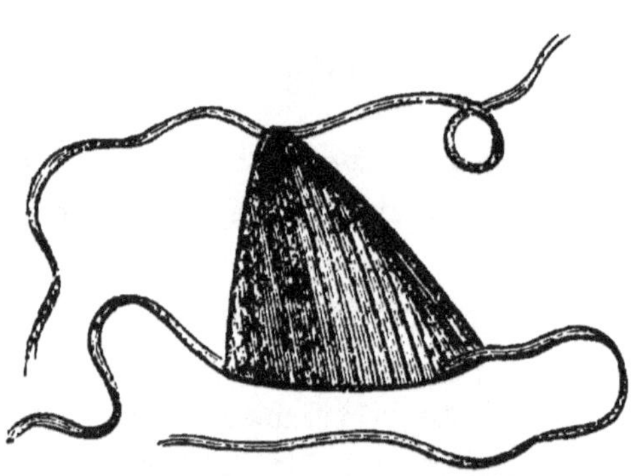

Fig. 33.—Stromeyer's Cushion.

tapes, passing in front of the chest and behind the back to the opposite shoulder, where they are tied ; the broad base is interposed between the chest wall

and the elbow, and thus forms a pad for the arm; it
is maintained in position by tapes passing round the
waist, and there tied.

Fractured Shaft of the Humerus.

Apparatus.—1. Three straight hollow splints. 2.
One L-shaped to reach from the axilla to the wrist.
3. Rollers 2 inches wide, and 1 inch for the fingers, or
straps and buckles. 4. Pads and wool. 5. Sling.

When broken below the attachment of the deltoid
and coraco-brachialis muscles the displacement of the
bone is commonly prevented with ease; neither shoulder
nor elbow-joint need be fixed, and it is not necessary
to apply the splints so tightly as to cause impediment
to the venous circulation. If, however, the pectoral
muscles or deltoid be connected with the lower fragment, the displacement is sometimes obstinate; in
such cases it is necessary to buckle the splints tightly.
For this to be safely done, the fingers and hand must
be previously bandaged to prevent œdema; with this
addition, the method of treatment is the same in both
varieties of fracture.

Step 1. Select the splints; they should be hollowed;
of wood, perforated sheet zinc, or wire gauze, about 2
inches broad, lightly padded, and provided with straps
and buckles.

The external one reaches from the acromion to the
outer condyle; the inner L-shaped one from the axilla
to the wrist; a third shorter one is placed behind the
arm, and if there is much projection forwards of the
lower fragment, a fourth very short one is added in
front. The patient should sit on a chair while the
apparatus is being put on.

E 2

Step 2. The fingers and thumb are bandaged, a little wool being placed in the palm of the hand ; the L-shaped splint, properly padded, especially at the upper end, which must not press up the armpit, is then applied along the inner surface of the arm and forearm, and maintained in position by an assistant, while a roller is applied round the hand and wrist, and up the forearm over the splint to the elbow, round which it is carried several times before being made fast.

The fingers need not be bandaged when the fragments can be kept in position without much pressure.

Step 3. An assistant grasping the elbow in one hand, pulls down the lower fragment, while he steadies the shoulder with the other. The displacement thus reduced, the surgeon applies the remaining splints, taking care that the inside splint does not reach too high into the axilla, lest it compress the axillary vein.

In simple cases the splints should be drawn close by straps and buckles ; where the muscles are powerful, a roller should be wound round the splints instead of straps.

Step 4. A 2-inch wide roller is fastened to the arm above the elbow, and then carried round the trunk to the arm again, to steady the limb against the body.

Fig. 34.—Fractured Shaft of the Humerus.

Step 5. The hand and wrist are supported by a

sling over the shoulders, the elbow being allowed to hang (see fig. 34).

This apparatus is worn three weeks, when the bandages are removed, and the splints applied less tightly than before. The inner L-shaped one may be replaced by a straight one reaching from the axilla to the inner condyle. All may be substituted by a sheath of gutta-percha moulded to the arm from the acromion to the elbow, and buckled on to the limb. The arm must be supported by splints of some kind for five weeks, but passive motion of the elbow and wrist should be adopted after the third week. The wrist especially should be set at liberty as soon as possible. In treating this fracture great care is necessary that the fragments be kept steadily in accurate position, as the humerus is prone to remain un-united for many months.

Fracture of the Anatomical or Surgical Neck of the Humerus, of the *Great Tuberosity*, and of the *Neck of the Scapula.* These fractures are similarly treated.

A. *By the shoulder cap. Apparatus.*—1. Paper for pattern. 2. Gutta-percha, leather, or mill-board. 3. Pads. A soft thin pad, 10 inches long, 5 inches wide (a treble fold of thick flannel or blanket answers very well), is wanted to line the axilla. If the cap is of leather or gutta-percha, a lining of wash-leather should be added after the splint is made. 4. Rollers, 2 inches and 1 inch wide for the fingers. 5. Scissors and pins. 6. A tray, and a kettle of hot water. 7. A towel, and basin of cold water. 8. Sling. 9. Cotton wool.

Step 1. If leather is used, cut out a paper pattern

of the splint on the limb to be fitted. The pattern should reach along the clavicle to the root of the neck, and over the scapula to its posterior border, and be continued down the arm to the elbow, tapering as it goes, but having its anterior and posterior margins meeting at the inner side of the arm to give the splint a good grasp of the limb. The end should be left long enough to turn a couple of inches round the point of the elbow (see fig. 35). A notch must be cut at the upper end of the paper pattern from the point of the shoulder to the root of the neck to make

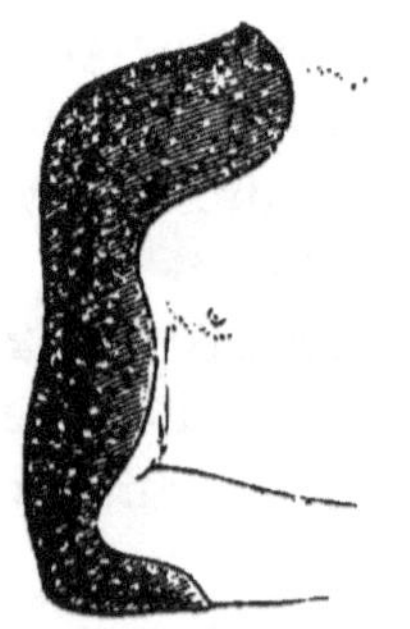

Fig. 35.—Cap for fracture near the shoulder.

it fit on the shoulder between the clavicle and the spine of the scapula. Two other notches must be cut opposite the folds of the armpit in front and behind to let the splint clip round the arm. These notches should be repeated in leather, if that is used, and the sides of the notch above the shoulder must be stitched together when the leather has set. Notches are unnecessary in gutta-percha, as that can be moulded on without them ; and for this reason the cap is much more serviceable when made of that material. The gutta-percha must be softened in the manner described in making the splint for the elbow at page 48 ; then accurately adjusted to the shoulder as high as the root of the neck, and to the arm at the axilla, and, lastly, turned under the point of the elbow a couple of inches (see fig. 35), while the fore-arm is well raised across the chest.

When set, the splint must be removed that it may be trimmed and lined with wash leather. If of gutta-percha, it must be perforated with small holes; if of leather, the notch at the shoulder must be stitched together. Next prepare a soft thin pad, 5 or 6 inches broad, and 8 or 10 inches long, to fill the axilla.

Step 2. Bandage the fingers and thumb separately, then, putting a little wool in the palm and round the wrist, bandage the hand and fore-arm as far as the elbow, where the bandage is fastened.

Step 3. Apply the splint. First get on the cap, drawing the fractured bone into posi-tion; then put the soft pad in the axilla (if the armpit is

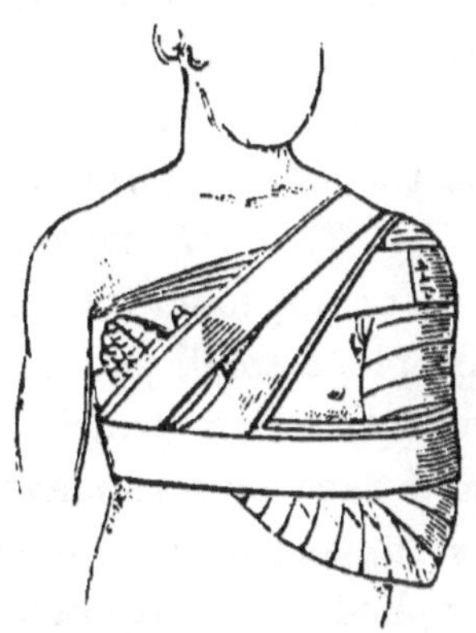

Fig, 36.—Fracture at the upper end of Humerus. The apparatus completed.

very hollow, fill it out with cotton wool), and bend the elbow till the hand lies on the breast of the op-posite side. Then, while an assistant holds the limb and apparatus in position, fasten them all in place by continuing the roller of the forearm in figures of 8 round the elbow until the splint is well fixed to it; and carry the roller up the arm by reverses to the axilla.

Step 4. A little wool or piece of flannel having been placed in the opposite armpit to prevent chafing, a spica for the shoulder is then applied (see page 17), but beginning at the root of the neck and working downwards to the point of the shoulder.

Careful extension is continued by the assistant all

the time this bandage is being put on, until the head of the bone is well drawn into the cap.

Step 5. The arm is drawn to the side, and the forearm fixed against the chest by a roller carried round the arm and trunk and over the shoulder (see fig. 36).

After three weeks the forearm may be released, but the cap and axillary pad must be continued to be worn two or three weeks longer while the arm is well drawn to the side, and the wrist carried in a sling.

B. *By simply confining the limb to the side.*

For impacted fracture of the neck occurring in old people, a small pad may be placed between the chest and the arm, which is then well drawn to the side and fixed by a bandage to the trunk, while the wrist is supported in a sling. Evaporating lotions can be applied to the exposed shoulder while the swelling and pain remain ; when these are gone, a starch bandage may be used until consolidation has formed.

In young children fracture about the head of the bone may be at once treated by applying a light, well starched bandage (see page 57).

Fracture of the great Tuberosity of the humerus is difficult to treat, on account of the tuberosity being carried backwards by the muscles and of the humerus being rotated forwards. Hence the parts must be braced together with a firm cap of gutta-percha moulded on to the shoulder while soft, and while the fractured parts are held in apposition, which may be done by the fingers, or by putting on a wet roller firmly over the splint as a shoulder spica until it is set. When the splint is hard the bandage may be taken off, and the splint removed and finished ready for application.   In doing this, the steps are the same as for fracture of the surgical neck of the humerus, and the necessity for fixing the arm well to the side of the body is as great as in that fracture.

Fracture of the Acromion is treated very much like fracture of the clavicle, that is, the arm is well raised by a sling under the elbow, and then fastened to the side.   It is not necessary to fill the axilla with a pad, as in fracture of the clavicle, for in this case the shoulder is not drawn inwards.

Fracture of the Clavicle.—*Apparatus.*—1. Axillary pad.   2. Roller, 3 inches wide.   3. Sling.   4. Wool.   5. Pins, or needle and thread.

Fractures of the clavicle nearly always leave some deformity after union; this is best minimised by keeping the patient on his back on a flat couch with the head alone raised by a cushion, and the arm fixed to the side until union has taken place.   As most persons will not submit to a fortnight's or three weeks' confinement in bed for this accident, the fragments must be kept in position as nearly as possible by apparatus while the patient goes about.

The displacement of the outer fragment is *inwards, downwards*, and *forwards*.   Many varieties of apparatus are employed to prevent this displacement during union; the following mode is perhaps as effectual as any other in accomplishing this object.

Step 1. Fix in the arm-pit a firm wedge-shaped pad of bedtick filled with chaff; 5 inches broad, 6 inches long, and 1½ or 2 inches thick at the thick end, or just big enough to fill the axilla and throw out the humerus without compressing the axillary vein, hence the thickness varies with the hollowness of the arm-pit (see fig. 37). A band and buckle are stitched to the thick end, which is uppermost. When in use, this

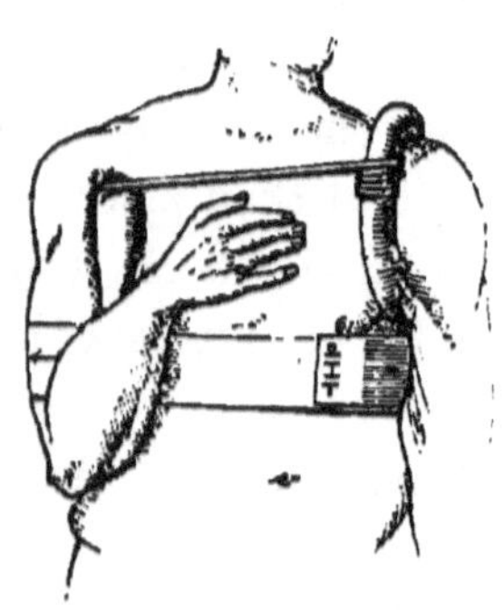

Fig. 37.—Wedge-shaped pad for broken right Collar-bone, attached to the American ring-pad.

band is passed over the opposite shoulder and keeps the pad in place. A little wool should be put under the band, where it crosses the root of the neck, to prevent chafing.

Step 2. The elbow is elevated by an assistant, who keeps the arm upright, and lays the fingers on the breast bone. A roller, attached to the arm by a couple of turns, is carried behind the back round the trunk, and over the arm above the elbow, drawing that close to the side.

Step 3. To support the elbow, the longest border or base of a three-cornered handkerchief is carried under it, one end passes in front, the other behind the body; both are then drawn tightly and crossed over the opposite shoulder, where one end is taken under the axilla, protected by a pad of wool, and the two are tied in front. In giving this direction the ring-pads shown in the figures are supposed not to be at hand. Lastly, the loose corner at the wrist is folded neatly and pinned up (see fig. 38).

This apparatus must be watched from time to time, and re-adjusted if any part slips. The sling and pad are to be worn for four weeks.

Union sometimes takes place in three weeks or less;

in which case the pad may be removed so much the earlier; but a sling should be worn for a fortnight after the bandage and pad are laid aside. In children the pad must be very much thinner and shorter than that described; the sling should be replaced by a bandage carried alternately round the body, and over the opposite shoulder. After it is put on the turns should be well stitched together and spread with stiff starch. In ban-

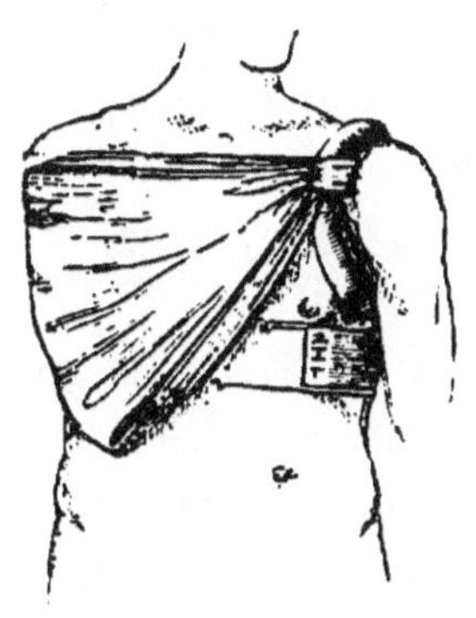

Fig. 38.—Apparatus for broken Clavicle finished.

daging children, great care must be taken to protect with wool the parts likely to be chafed.

The American surgeons have a very good plan for attaching the sling to the sound shoulder. Instead of carrying the ends of the sling round the sound shoulder and under the axilla, they pass over the shoulder a loose but well-stuffed collar or ring-pad (see fig. 37), to which they fasten the ends of the sling in front and behind; this prevents all cutting or chafing under the armpit, and distributes the strain evenly.

Sayre's Method.—Sayre, of New York, recommends a mode which is frequently adopted in University College Hospital :—

Step 1. A small folded compress is passed up the armpit to fill the axilla.

Step 2. One end of a strip of stout plaster 4 inches wide, and long enough to wrap round both trunk and arm, is looped round the arm, plaster side outwards, and made fast by stitching the end down behind the limb. The plaster is then heated by running a hot iron, or can of hot water, along the back of the strip ; and, while the arm is drawn downwards, the plaster is carried closely behind the trunk and over the arm to the back again, where this second end may be also secured by a stitch. This holds the arm closely to the trunk.

Step 3. A second strip of plaster, of width and length equal to the first, is fixed by one end to the acromion process on the sound side. It is then brought down across the back to the elbow, whence it is drawn tightly upwards across the front of the chest over the arm to the sound acromion again, thus fixing the injured arm and thrusting upwards the shoulder. To make them fast, the ends of this strip should be sewed together.

Figure-of-8 bandage.—Many surgeons still employ a figure-of-8 bandage carried under each armpit and crossed behind the back. Under any circumstances this plan is irksome to the patient, but is least so if two silk handkerchiefs be substituted for the bandage, one being passed round each shoulder and the ends of both braced tightly together behind the back. A little wadding should be rolled up in the centre of each silk handkerchief, to prevent the arm-pit being galled. The wedge-shaped pad may be dispensed with if the shoulders are braced back, but the elbow must still be raised and drawn to the side.

LOWER EXTREMITY.

A ruptured tendo Achillis is treated by extending the foot and flexing the knee ; for this purpose the patient wears a high-heeled slipper. A band is sewn to the heel, drawn tight, and fastened to a buckle and strap round the thigh, just above the knee. The patient should not walk for a month unless he will use a wooden leg on which he can kneel, with the knee bent.

Separation of the Epiphysis of the Calcaneum, which sometimes occurs instead of rupture of the tendo Achillis, is treated in the same way.

A more secure method of maintaining extension of the foot is to apply a narrow, straight, well-padded splint, along the shin to the toes, and to bandage the leg and foot firmly to it.

Fracture of the Fibula.—*Dupuytren's Splint.*— When the fibula only is broken, it may be treated in several ways ; this, however, is the common plan :—

Apparatus.—1. Straight wooden splint. 2. Pad and wool. 3. A roller. 4. Pins.

Step 1. The splint should be about 3 inches broad, and long enough to reach from the head of the tibia to 4 inches beyond the sole of the foot. A notch $1\frac{1}{2}$ or 2 inches deep is cut at the lower end of the splint in which to catch the bandage. The splint is then padded, care being taken that the padding is sufficiently thick to prevent galling at the upper end against the inner tuberosity of the tibia, and also that it becomes thicker as it descends along the leg, for that part to rest easily against the splint ; lastly, the pad should end in a thick boss or projection opposite the internal malleolus, beyond which it should not reach, lest it interfere with the rotation and adduction of the foot inwards.

Step 2. The splint, when thus prepared, is applied along the inner side of the leg, care being taken in doing this that the internal malleolus is against the middle of the splint and not allowed by the assistant to slip towards the anterior or posterior border.

Step 3. A roller is then carried round the limb and splint, beginning below the knee and continuing in simple spirals for three or four turns, when it is fastened and cut off.

Step 4. A light layer of wool is wrapped round the outside of the ankle, heel, and back of the foot. Then a roller, beginning at the splint, passes outwards in front of the ankle over the external malleolus, behind the heel to the splint; then over the ankle to the outer margin of the foot, next under the sole through the notch of the splint to the front of the ankle joint again, where it repeats the same course three or four times. Each turn must be tightly applied and made to draw the foot well inwards to the splint, and in doing so to tilt outwards the broken part of the fibula (see fig. 39).

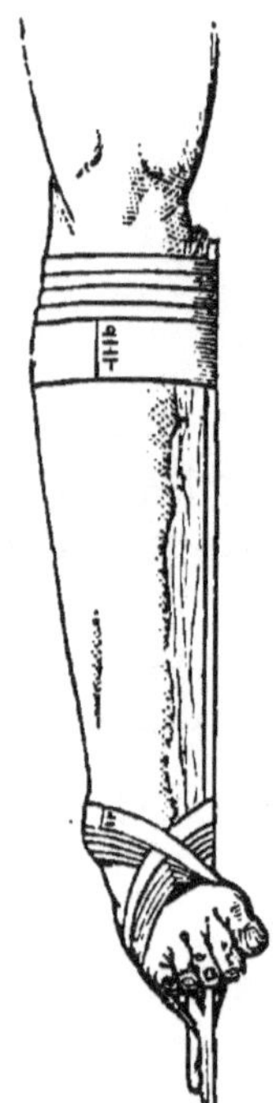

Fig. 39.—Dupuytren's Splint for fracture of the Fibula.

This splint is cumbersome, hence is by some surgeons replaced by a light starch or gum casing for the foot and leg, leaving the knee free. Croft's plaster of Paris splints are well suited for this and the next fracture, see p. 95.

Fractures of the Tibia with or without breakage of the fibula, and fractures at the ankle joint.

These fractures are often, from their obliquity, difficult to keep in good position ; in such cases *McIntyre's Splint* (fig. 40) is very generally used in the early part of the treatment. For this splint the following *apparatus* is required :—1. McIntyre's splint. 2. Pads for the double incline plane and foot-piece. 3. Sock of flannel for the foot. 4. Rollers, 3 inches wide. 5. Wool, pins, needle and thread, strapping plaster. 6a. A sling cradle ; or, 6b. Board, block, gimlet, screws and screw-driver.

The McIntyre's Splint may be used either bent or straight ; in whichever position of the knee most relaxes the tension of the muscles on the fragments. As a general rule the straight position is best if the fracture is high up, and the bent one when the bones are broken near the ankle joint.

Step 1. A splint of suitable length is selected, by measuring the sound leg. It should be no wider than the limb. The joint of the splint should be put behind the knee, and space be left beyond the foot for the foot-piece to slide along the slots when extension is made.

The splint is next padded, the hollow where the lower part of the calf and small of the leg will come being well filled, that the leg may be thoroughly supported ; but the space behind the heel and tendo Achillis must be left quite clear. A small pad is then fastened by a strip of strapping or by needle and thread to the foot-piece.

Step 2. The limb having been first cleaned and dried, the foot and ankle are wrapped in an even layer of cotton wool. A sock or boot made of flannel is next put on the foot. This may be readily extemporised by cutting off the foot of an angola stocking, slitting it up along the back to the toes, and sewing on to the sole, one inch in front of the heel, the middle of a piece of tape $\frac{1}{4}$ inch wide and 18 inches long. The foot is then wrapped in the sock, the edges of which are drawn together by a needle and thread, care being taken that the sock fits closely round the ankle and back of the foot. A little wool having been wrapped round the knee, the limb is next raised while the splint is placed under it ; the screw is turned until the inclined planes are at an angle suited for the maintenance of the fragments in position, and the foot-piece is pushed up to the foot with its screw-pin loose, that it may be adjusted to the amount of flexion or extension necessary for the foot ; this being ascertained, the screw is tightened to keep it so while the foot is fastened to the foot-piece. For this the strings of the sock are brought over the top of the foot-piece, and drawn tight before tying them.

The position of the heel is very important. It should not sink below the splint, or it will rest on the bandage ; neither should it be drawn up too high, or the weight of the leg will hang on the sock, instead of resting on the pad ; both frequent causes of pain at the heel. When the proper position is obtained, the strings are made fast to the pin behind the foot-piece, and the foot is steadied by two or three turns of a roller carried round it and the foot-piece (fig. 40).

Step 3. The thigh is next fastened to the thigh-piece by a roller carried from the top of the splint downwards along the thigh to the knee, or even below that joint if the fracture is near the ankle.

In doing this the roller is passed inside the screw, should that be placed underneath the splint, as in fig. 42, p. 67, and not at the side as in fig. 40—for the screw will be wanted free for further adjustment.

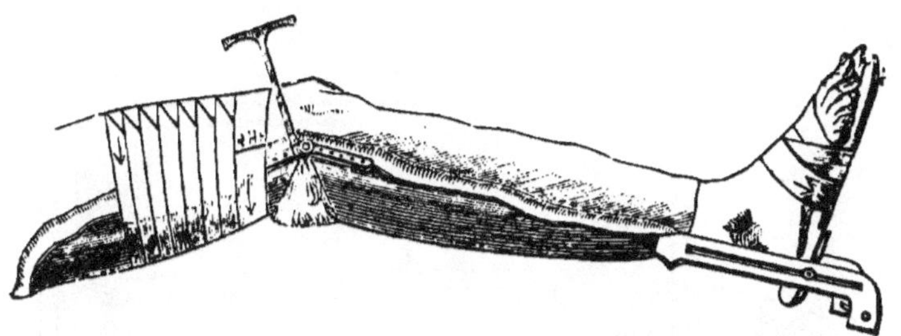

Fig. 40.—McIntyre's Splint. The thigh fixed ready for extension
of the leg.

Step 4. An assistant grasps with both hands the foot and foot-piece, and pulls them downwards until the shortening is removed. While doing this, he tilts the foot up or down as the surgeon finds necessary for adjusting the fragments ; the knee and the leg are also raised or lowered until a good position is attained. The general rule is to keep the great toe and the inner malleolus in a line with the inner border of the patella. This done, the surgeon tightens up the screw-pin of the foot-piece, and completes the attachment of the foot by continuing his roller with figures of 8 round the foot and ankle ; these turns should not however pass above the fracture, and should be no more than sufficient to secure the position of the foot and of the lower fragments (see fig. 42).

Step 5. The bandaging usually ceases here. If the limb swell, a separate roller may be carried along the leg to support the muscles and restrain œdema, otherwise the leg is best left bare, that the position of the fragments may be watched, and evaporating lotions applied.

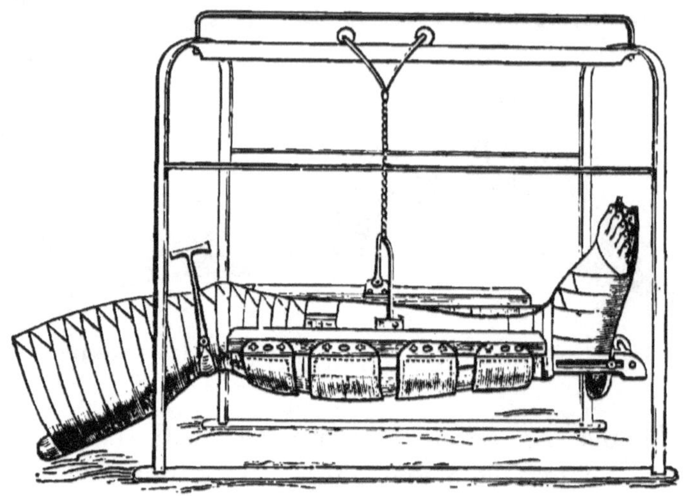

Fig. 41.—McIntyre's Splint slung in Salter's Cradle.

Step 6*a*. This consists in slinging the limb, for which Salter's Cradle is very convenient (see fig. 41), or an ordinary bed cradle answers very well, from which the limb can be slung on pieces of bandage carried underneath the splint at the knee and ankle.

6*b*. Instead of elevating the limb by a sling, it is also customary to raise and fix the splint on a block (fig. 42), 6, 8, 10, or even 12 inches high, as may be necessary; this block slides in a groove on a board 3 feet square, put between the mattress and bedstead, to afford a firm support for the block.

In ordinary cases the limb is kept on the splint three weeks, until the irritation has subsided, and partial union is attained; the splint may then be replaced by a starch bandage, and the patient may soon leave his bed.

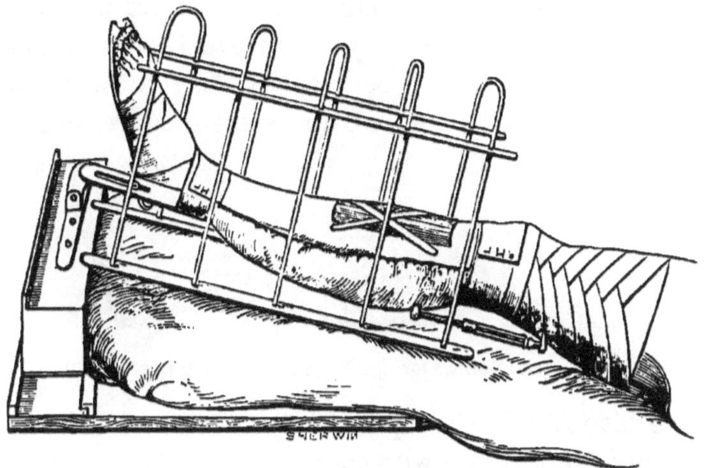

Fig. 42.—McIntyre's Splint raised on a Block.

Transverse Fracture of the *tibia* alone, or even of both bones, when the displacement is small, is very well treated by a hollow splint on each side. Both splints are cut away opposite the malleoli, and the inside one may end at the tarsus, though it is better if, like the outside one, it reaches to the toes. The splints reach on each side to the head of the tibia, but ought not to extend above the knee-joint (see fig. 43).

Step 1. They are padded lightly and evenly along their whole length and applied to the limb on each side.

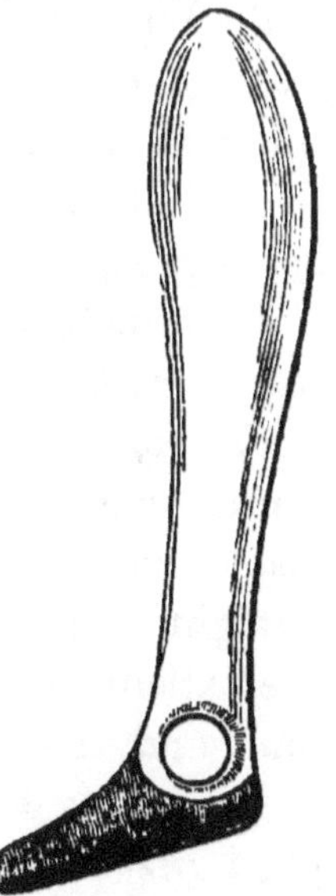

Fig. 43.-Outside lateral Splint for the Tibia.

Step 2. They should then be secured by figures of 8 round the foot and ankle until the foot is firmly fixed in them. The bandage should then be fastened off, and extension made by an assistant, who grasps the foot and ankle with both hands while the surgeon fixes the splints to the limb above the fracture, begin-

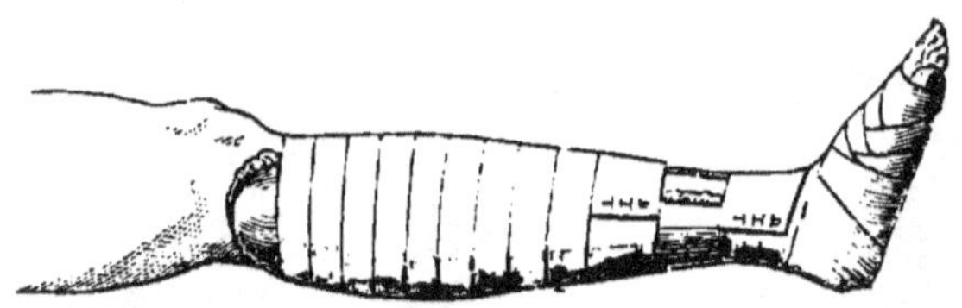

Fig. 44.—Lateral Splints for simple transverse fracture of the Tibia.

ning his roller at the top just below the knee, and continuing it downwards with spiral turns until the fracture is reached, above which point it should terminate (fig. 44). When the apparatus has been applied, the limb may be either supported upright by sand-bags, or slung in a cradle for three weeks, after which time the splints are advantageously replaced by a starch bandage for three weeks more.

For a method of applying plaster of Paris splints to recent fractures of the tibia, see page 95.

Flexing the Leg for Fracture of the Tibia.—Sometimes, when there is unusual difficulty in preventing displacement of the fragments while the limb is nearly straight, the bones can be readily kept in position if the patient lie on the same side as that of the injured limb and the knee be well flexed. For such cases these splints are very suitable; they should be applied *after* the limb has been bent and the fragments brought into apposition. When the splints have been put on, a roller may be carried round the leg and thigh to keep the knee in its bent position.

The Horseshoe Anterior Splint may be used when there is much displacement of the foot backwards, which cannot be overcome by suspending the limb by the foot.

Apparatus.—1. Straight wooden splint of the shape described below. 2. Pad. 3. Roller, $2\frac{1}{2}$ inches wide. 4. Two handkerchiefs. 5. Cotton wool. 6. Two strips of plaster. 7. Pins, needle, and thread.

Step 1. The splint is 3 inches wide at the top, and, tapering slightly, reaches from the tubercle of the tibia to the front of the ankle, where it widens again rapidly, and bifurcates into two horns, 6 inches long; if hollowed it sits better on the leg. Two holes, $\frac{1}{2}$ an inch wide, are made at the upper end of the splint (see fig. 45). The pad should be made thicker at the sides than in the middle (and most thick on the inside), to preserve the edge of the tibia from pressure. The pad should also be considerably thicker at the lower part, that the splint may rest evenly on the anterior surface of the limb. The pad and splint are fastened together by strips of plaster.

Step 2. The splint, thus prepared, is laid along the front of the leg, while an assistant maintains the foot in good position.

Step 3. A folded handkerchief is next passed round the back of the limb at the top of the splint, and the ends, brought through the holes, are tied in front.

Step 4. A roller is applied from above downwards, and fastened off at the ankle.

Step 5. A 'bird's nest' pad of cotton wool is laid under the point of the heel. A folded handkerchief is next carried beneath the heel and wool, one end being brought up on each side of the foot, when a turn is taken round the horn of the splint, and both

Fig. 45.—Horseshoe Anterior Splint.

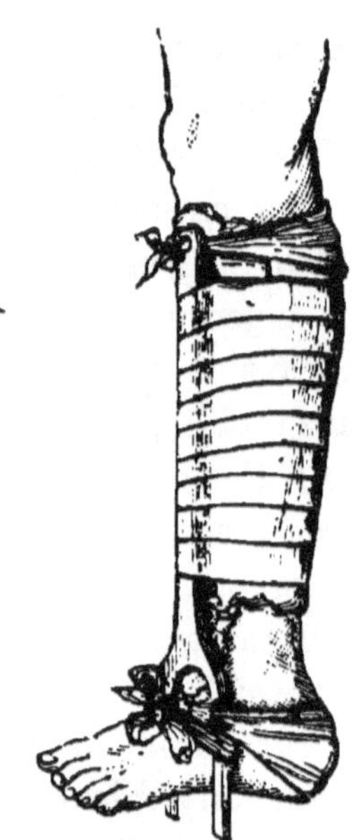

Fig. 46.—Anterior Splint applied.

ends knotted together over the foot (see fig. 46). It is a good plan to nip up a small fold of the handkerchief, before and behind the heel, with a needle and thread.

Fracture of the Patella.—When this bone is broken there is usually much swelling from effusion into the knee-joint; while this is present, rest, with cold lotions, and elevation of the foot, are generally considered necessary. When the effusion has subsided, the upper fragment must be brought down to the lower one, by some means like the following.

Apparatus.—1. Straight wooden splint with a foot-piece. 2. Pads. 3. Diachylon plaster. 4. Roller. 5. Lint and wool, pins. 6. Two hooks or screws, gimlet, and screw-driver.

Step 1. The splint is first fitted; it should reach from the buttock to the heel, at which point a foot-piece rises for the foot to rest against; at the back of the splint a line should be marked 3 inches above, and another 3 inches below the knee-cap, into which a stout screw or hook is inserted before the splint is put on. It is then well padded, to support the ham and leg, while the heel is left free, and a pad is put between the sole and the foot-piece. A firm crescent-shaped pad is prepared to sit like a saddle above the upper fragment.

Step 2. The limb is laid on the splint, while an assistant draws the patella as nearly as possible into its place; the surgeon lays the crescentic pad on the thigh above the patella. He next takes a strap of plaster 2 inches broad and 20 long, warms it, and lays the middle across the compress, drawing each end first tightly round the limb, and then downwards and forwards in a figure of 8; a similar strap is fixed below the lower fragment. The knee, shin, ankle, and foot are then protected by a layer of cotton wool, and the bandaging begins.

Step 3. The roller first fastens the foot against the foot-piece by figures of 8, then passes up the leg by reverses until it arrives opposite to the lower hook, where it is fastened.

Step 4. A second roller is then begun at the top of the thigh and brought down the limb till it reaches the compress above the patella; from this point it passes below the lower screw at the back of the splint and makes one circular turn round the leg; the roller is then taken upwards across the compress (as shown in fig. 47) to the upper screw, where it also makes a circular turn; having done this it again descends to reach the lower screw, and is returned as before. Each of these turns should be drawn tightly to bring the upper fragment as near the lower one as possible;

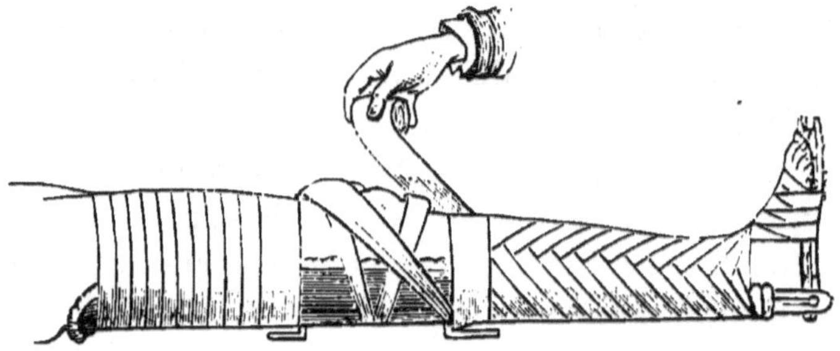

Fig. 47.—Fractured Patella, drawing down the upper fragment.

when this is done the bandage is completed over the knee by figures of 8. It suffices to fix the lower fragment, which cannot be drawn up to meet the upper one; the latter must descend to it.

Step 5. The limb is lastly put into position by elevating the heel and by raising the body with pillows till it is in a half-sitting position.

The patient wears this splint for four weeks, during the first fortnight of which the bandage should be perseveringly re-applied every three or four days until the upper fragment is brought into apposition with the lower one. After four weeks the splint may be changed for a light starch or gutta-percha case, to be worn for six weeks more, and then replaced by a back splint of leather and knee-cap, that must not be laid aside for another period of four weeks.

If the patient can be persuaded not to bend his knee for six months, the union of the fragments will be less likely to yield afterwards. He should be also warned that much stiffness will result from the long-fixed position necessary to procure good union between the fragments : but the stiffness will all pass away in time, notwithstanding the long-enforced rigidity.

Another Plan of treating fractured Knee-cap consists in applying a *well-starched apparatus* as early as possible after the accident. *Apparatus.*—1. Sheet of millboard. 2. Rollers. 3. Cotton wool. 4. A basin of freshly-scalded starch. 5. Lint and strip of plaster.

While the apparatus is being put on, the knee must be kept perfectly straight.

Step 1. Prepare two pasteboard splints, to reach from just below the groin to within 3 inches of the ankle joint, wide enough to nearly enclose the limb, leaving an interval of 2 or 3 inches between the edges in front.

Step 2. Envelop the limb in a thick, even layer of cotton wool. If there be much difficulty in maintaining the fragments in good apposition, place a small pad of lint above the upper one, and fix it by a strip of plaster, carried in a figure of 8 underneath the knee to the front of the leg.

Step 3. Next bandage the foot and ankle ; apply the splints prepared as directed at page 87, and carry a roller, well starched as it is laid on, from the ankle to the groin; care should be taken to cover in the knee well by figures of 8, arranged to fix the upper fragment securely.

Step 4. After applying one or two more layers of starched bandage, put the limb into position by raising the heel till the starch is dried.

If the case, when dry, is at all loose, it should be cut open. When the position of the fragments has been ascertained to be satisfactory, the edges of the case must be pared where they overlap and re-adjusted by a tightly-applied bandage.

The patient may get about on crutches in eight or ten days' time.

Fracture of the Shaft of the Thigh-bone.— *The long Splint. Apparatus.*—1. A wooden splint. 2. Rollers, 3 inches wide (one of 6 inches wide). 3. Perineal band. 4. Strapping, needle, thread, and pins. 5. Pad and wool.

The splint for an adult should be $2\frac{1}{2}$ or 3 inches wide, and long enough to reach from the nipple to 6 inches beyond the heel; two round holes, of $\frac{3}{4}$ inch diameter, are cut at its upper end, and at the lower end two notches 2 inches deep.

Liston's Mode of applying the Long Splint.—Step 1. The limb is first washed with soap and water, well dried, and afterwards dusted with starch powder, especially at the perinæum.

Step 2. The end of a roller is split for a few inches, and tied in the holes at the upper end of the splint. The roller itself is carried down the inside of the splint and attached temporarily to the notches at the other end; a pad is then fastened on, by drawing the margins together with needle and thread across the outside of the splint, or by tying strips of bandage round the pad and splint at short distances.

Step 3. Prepare the perineal band. This consists of a silk handkerchief or napkin folded into a *flat* ribbon, 1 inch wide and covered for about 1 foot of its length with oiled silk. A piece of smooth brown paper, 1 foot long and 4 inches wide, folded into a ribbon 1 inch wide, makes an excellent foundation for the silk handkerchief to be folded upon. A band thus prepared is too stiff to become a cord after it has been worn a few days, which a simple handkerchief is apt to do. One end of the band is passed in front of the groin, and one behind the buttock, great care being taken that it presses on the tuber ischii in the perinæum, and not between that point and the lesser trochanter.

Step 4. The ankle and back of the foot are wrapped in a layer of cotton wool, and the splint applied along the outside of the body. The ends of the perineal band are drawn separately through the holes in the splint, and left loose. The bandage which was fastened to the splint is now released from the notch; and, taking with it the end of the pad, is carried under the sole, then in front of the ankle to the splint, and behind the leg round the internal malleolus to the front again. Here it crosses outwards and goes thence through the lower notch of the splint to the inside of the foot again. This figure of 8 is carried four times over the dorsum of the foot, twice through each notch of the splint, and is made fast by a pin or a stitch. In doing this, care must be taken to keep the leg and splint parallel, and that the splint does not ride over the back of the foot; the external malleolus should be midway between the margins of the splint; moreover, the bandage must fit firmly round the ankle and splint, spreading over the dorsum no more than can be helped, to avoid straining the front of the ankle. (Means for more effectually preventing this strain will be afterwards detailed, on p. 77). All being ready for extension, an assistant, grasping the leg and splint above the ankle, pulls out the shortening till the broken bone is in a good position, while the surgeon tightens the perineal band, and makes the ends fast in a knot.

Step 5. The surgeon returns to the foot of the patient, and having laid some cotton wool along the shin and round the knee, continues the bandage. This is carried up the leg and over the knee by reverses and figures of 8, as high as the seat of fracture. It is customary to carry the bandage up the thigh; but this addition is not an essential part of the apparatus, which is simply to keep up the extension in the direction of the axis of the limb. A bandage conceals the limb and the position of the broken ends of the bone ; but it steadies the thigh and confines the muscles, thereby often preventing pain.

Step 6. First protecting the bony parts with cotton wool, the muscles about the hip are confined by a spica carried round the body and the splint, not merely a simple figure of 8 as depicted in figure 50, but a series of overlapping turns which ascend and cover in the hips well. Afterwards the upper end of

the splint is drawn close to the body by a few turns of
a broad roller carried round the *chest* (not the waist)
from above downwards (see fig. 50).

The perineal band must be changed whenever it gets
soiled, and the skin washed before a clean one is
applied. After the first few days the band need not
be very tight; it suffices if not slack or loose. Mr.
Coxeter makes india-rubber tubes in the shape of a
perineal band; these are filled with water when in
use (see fig. 48).

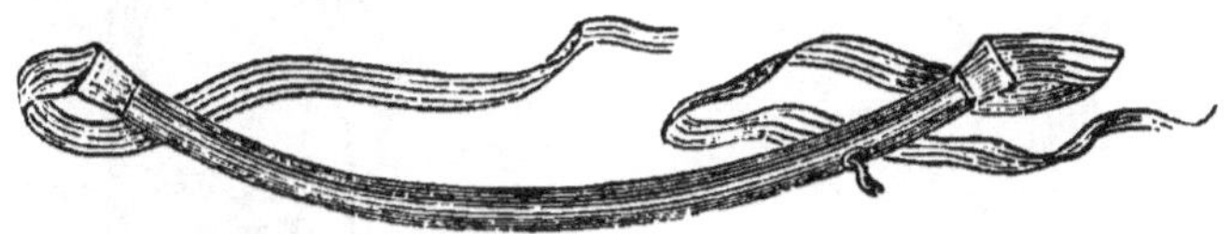

Fig. 48.—Coxeter's elastic perineal band.

Stirrup extension is a mode of relieving the strain on
the front of the ankle, caused by the pressure of the
roller which affixes the splint to it. A 3-inch wide
roller or bit of wood of the same breadth is laid against
the sole of the foot, and a stout india-rubber ring
2 inches in diameter is slipped over it. A piece of
strapping plaster, $2\frac{1}{2}$ feet long and 2 inches wide, is
passed half-way through the ring, and fixed securely to
the bandage or bit of wood forming the bar of the
stirrup, after which the ends of the plaster are carried
up the leg on each side. The plaster is kept in place
by a second strip laid on in spirals up the limb as
in fig. 49, and the india-rubber ring is hitched against
a hook at the end of the splint. By this means the
strain is transferred to the leg, and the ankle is left

free.   It is perfectly successful, and very easy to the
patient.

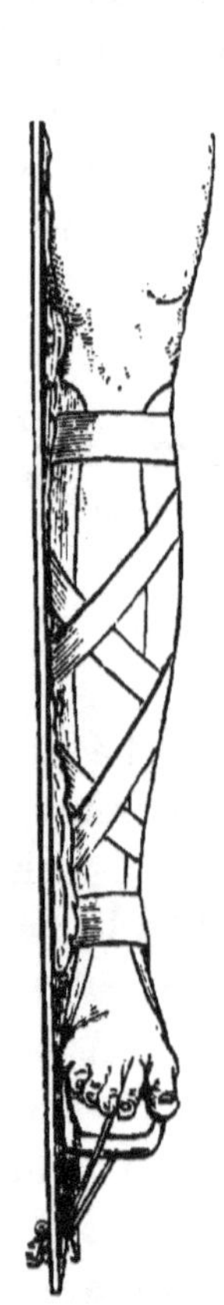

Fig. 49.—Mode of fasten-
ing the stirrup to the
leg, to avoid straining
the ankle.

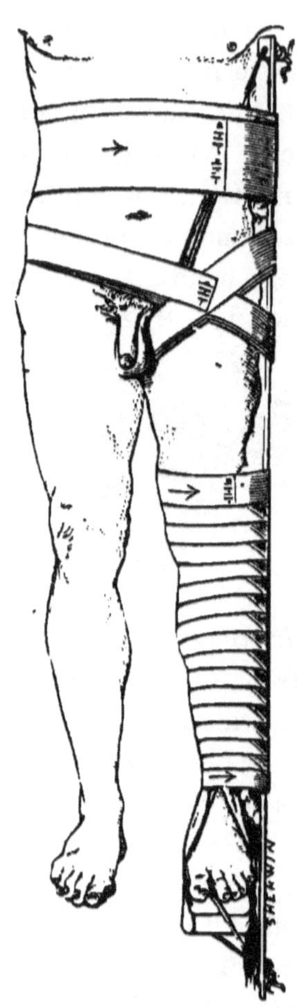

Fig. 50.—The long splint, with
elastic stirrup extension at
the foot. Bandage carried up
to the seat of fracture.

Mr. Buckston Brown has improved the mode of
attaching the elastic ring to the foot of the long splint;

he uses a hook passing through a small bracket, and having a nut with a screw-thread, in which the end of the hook runs. Hook and bracket are fastened to a block of wood, which, in its turn, can be screwed to the end of the splint. By drawing the hook downwards and tightening the nut the proper amount of yielding strain can be obtained (see fig. 51). There are several other ways of maintaining a continuous but slightly yielding strain on the muscles of the thigh.

The long splint is to be worn continuously for six weeks : or, what is better, after the first three weeks it may be replaced by a starch bandage, and the patient allowed to get about on crutches with his leg slung from his neck.

*The Scotch method of applying the Long Splint.—Apparatus.—*1. A long splint as already described on page 74. 2. A sheet of millboard for short splints. 3. A sheet or tablecloth. 4. Two large-sized handkerchiefs or slings. 5. Cotton wool, oiled silk, and bandage. 6. One dozen stout carpet pins.

Fig. 51.—Buckston Brown's thigh extension.

Step 1. Prepare two pasteboard or thin wooden splints, one to reach from the trochanter major to the external condyle, the other to extend from the inner side of the groin to the internal condyle ; pad the splints lightly with cotton wool. If required, a short anterior splint may be added, as seen in the figure.

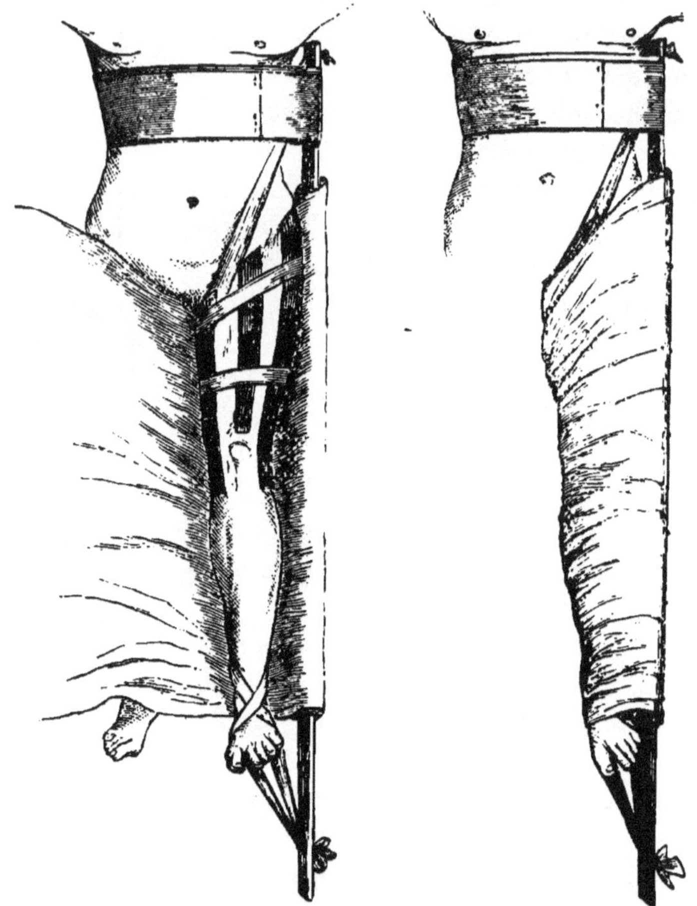

Fig. 52.—To apply the Scotch Sheet.   Fig. 53.—The Scotch
                                                     Sheet finished.

Step 2. Fold the sheet so that it will extend from the heel to the crest of the hip, and wrap it round the long splint, leaving unwrapped only sufficient to envelop the limb.

Step 3. Next place a folded handkerchief round the back of the leg above the ankle, and cross the ends over the foot. Another handkerchief should be applied as a perineal band (see page 75).

Step 4. While one assistant holds the ends of the perineal band, another grasps the leg above the ankle and extends the limb. The surgeon then applies the short thigh splints, and firmly secures them by loops of bandage.

Step 5. He next adjusts the long splint to the outer side of the limb, passing the free portion of the sheet under the ankle, leg, and thigh, and folding down the upper border of the sheet sufficiently to allow it to fit evenly at the perinæum and over the groin in front. The ends of the perineal band are then passed through the holes in the splint and tied firmly together, and the ends of the lower handkerchief are also fastened, by tying them over the notch, below the foot. Thus extension is maintained.

Step 6. Having protected the point of the heel with a ' bird's-nest ' of cotton wool, the surgeon draws the sheet smoothly and evenly over the front of the limb, and fastens it by carpet pins along the anterior edge of the splint.

Step 7. Finally, the upper end of the splint is maintained in position by a few turns of a chest roller.

In children fractures of the femur (and oftentimes hip disease) are best treated by Hamilton's splint (see fig. 54).

Two side splints connected by a cross-piece at the lower ends, long enough to reach upwards, nearly to the armpits, but separated a little more widely below than above, to render the perinæum more accessible, are laid on each side of the body. Strips of millboard are laid round the broken thigh, and the injured limb is firmly bandaged to the splint. The coaptation splints, the trunk, and the sound limb are all made fast by strips of stout plaster. Counter-extension is seldom needed in children, because the periosteum is rarely torn through. But if there be much shortening, a perineal band can be added. It is nevertheless

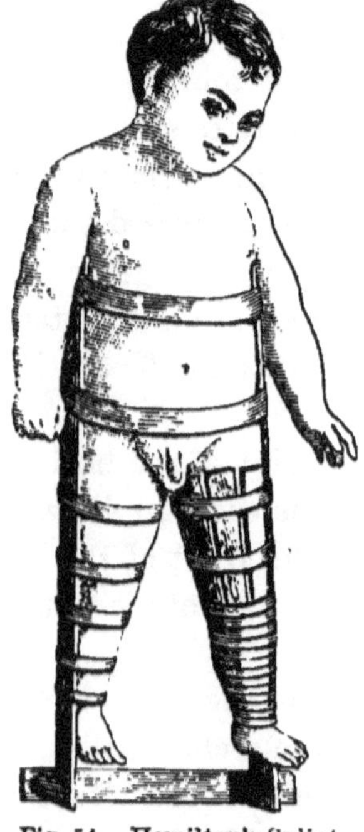

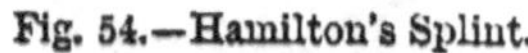

Fig. 54.—Hamilton's Splint.

Fig. 55. — Fracture below the trochanters ; bone in angular union.

troublesome, as it is constantly soiled. A child so fixed can be lifted from the bed or turned over without risk of displacing the broken ends of the bone.

Continuous Extension with the Limb flexed.—

The muscles attached to the upper end of the femur sometimes cause so much flexion and rotation outwards of the upper fragment that union of the bones in such a position produces a result which is illustrated by fig. 55, drawn from a preparation in the museum of University College.

This crooked union is prevented by bending the thigh and relaxing the muscles of the hip. This object is accomplished by using the *double incline planes*, as shown in figs. 56 and 57.

The limb is raised over a wooden frame abou 8 inches broad, with a double slope high enough at

Fig. 56.—Double incline planes.

the apex for the leg and foot to hang unsupported down the further side (fig. 56). It is well padded before being applied, and the leg and thigh are secured to it by a roller passed round the limb and plane. A better mode of steadying the limb is to fix a trough of gutta-percha while the limb lies on the plane. When the trough is set, it is screwed down to the wood at one or two points.

Slinging the double incline planes was practised many years ago by Mayor of Lausanne, and has been improved by Nathan R. Smith and Hodgen of America. It is an apparatus very easy for the patient, and particularly well suited for compound fractures of the thigh, for fractures near the trochanters that require a flexed position, or for fractures of the neck of the femur where the patient's feebleness does not permit the constraint of the long splint.

Apparatus for Hodgen's Method.—1. A bent wire frame (see fig. 57) with a separate foot-piece.   2. Two

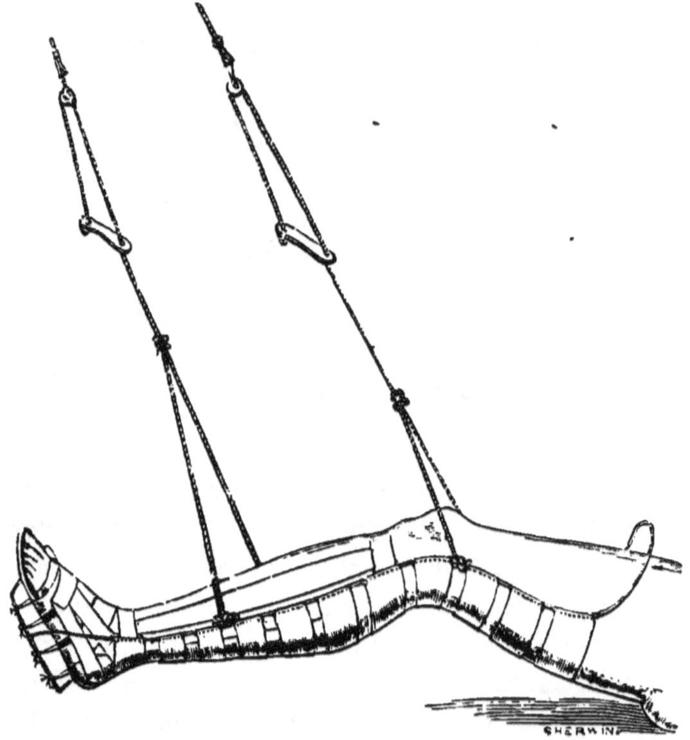

Fig. 57.—Double incline plane, slung.

pulleys, a rope with tent stretchers passing up to hooks in the ceiling, or to some suitable support.   3. One long and one short soft pad.   4. Strapping plaster, and some ends of bandage.

Step 1. The limb is washed and dried, and the short pad fitted to the foot-piece, which is furnished with some hooks at its lower surface, where ends of bandage or tape can be fastened, for fixing it to the wire frame. The frame is next prepared by passing strips of bandage across it from side to side at short intervals, to make a support on which the limb is laid ; if there is no wound, a soft pad may be put on the frame first, but if one be present, the limb should rest immediately on the strips of bandage, which can be changed whenever soiled, and replaced by clean ones without disturbing the limb. These strips should be tacked on with a needle and thread, so that, when the limb is placed on the apparatus, they can be shortened or lengthened till the leg bears evenly on them.

Step 2. The foot-piece is adjusted and fastened to the foot by straps of plaster carried round it and up each side of the leg, as was done for the stirrup extension in the 'long splint' (p. 77).

Step 3. The limb is next placed in the cradle formed for it, to the lower end of which the foot-piece is tied securely ; the ropes are rove through the pulleys and tightened till the limb swings easily. The point of attachment of the ropes must not be just above the limb, but beyond it, that the leg may be drawn away from the body along its own axis. The weight of the body makes counter-extension sufficient to remove all shortening in a few days.

Fracture of the shaft of the thigh bone in the adult may also be treated by simple extension, with a weight and pulley, as is described on the next page, for hip disease or fracture occurring in young children. It is a good plan in most cases, to enclose the thigh within three or four short well-padded wooden splints, as they serve to maintain the ends of the bone in good position.

Continuous Extension with the Knee straight is employed for fractures of the femur and in hip-disease. It is procured as follows : A stirrup is fastened

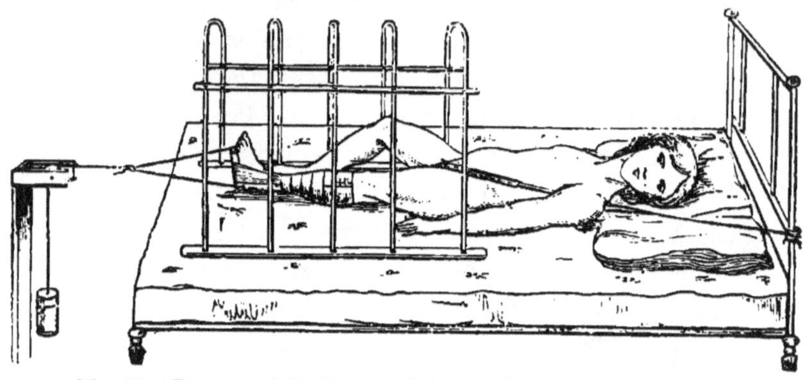

Fig. 58.—Fracture of the femur. Extension by weight and pulley.

to the leg in the way described at p. 77 ; to this a cord and weight are attached below the sole of the foot, and passed over a pulley fixed to a tripod frame (fig. 58), or any convenient object beyond the bed, in a line with the axis of the limb. The weight should balance the contraction of the muscles, and usually varies between 2 and 6 lbs. A perineal band fastened behind the patient's head keeps the body from following the limb. The weight may be a common scale weight, or a bag with a hole at the bottom closed by a string, and filled with shot or sand, or a can with a tap at the bottom filled with water : these arrangements allow increase or lessening of the weight, without slackening the cord and moving the limb. This apparatus requires no bandages, which are difficult to keep clean in children, and exerts a very even and continuous strain on the limb.

The pulley-stand should be capable of being made taller or shorter, so that in cases of hip disease with flexion, extension can be made while the thigh is still bent at the hip. By lowering the standard as the flexion gives way, the plane of extension can be slowly brought down to the plane in which the trunk lies. The trunk should always lie flat on the mattress, so that the spines of the lumbar vertebræ are not arched upwards. When they assume that position the pelvis is tilted forwards, and the plane along which extension of the thigh is made is too near to the horizontal.

The perineal band may be often dispensed with, by laying the patient on a flat mattress and raising the foot of the bedstead a few inches higher than the head; the body then sinks towards the head of the bed and resists the extension of the leg.

Starch bandage.—The following mode of applying the starch bandage and pasteboard splints may be used in all varieties of fracture; the length of the splints and the number of joints that should be included depend on the bone that is broken.

Some surgeons apply the starch apparatus immediately after the fracture has happened, others wait until partial union is procured and the irritability of the muscles and swelling of the limb have subsided.

Apparatus.—1. Sheets of bookbinder's millboard. 2. Rollers suitable for the size of the limb. 3. Cotton wool. 4. A basin of freshly scalded starch. 5. A long strip of plaster, to reach as high as the bandage will extend up the limb. 6. If the fracture be recent, a wooden splint will generally be necessary to keep up extension while the starch is drying.

As a general rule, the joint at the lower end of the fractured bone should always be fixed, and that at the upper end also, if the fracture is near that point. For an example of the mode of fitting, let us suppose the femur to be broken between the middle and lower thirds as in fig. 60.

Step 1. The limb is first measured for the splints. The length from the top of the sacrum to the heel, from the tuber ischii to the inside of the foot, and from the iliac crest to the outside of the foot, should be taken, and three strips of millboard prepared of corresponding lengths; the posterior one being 3

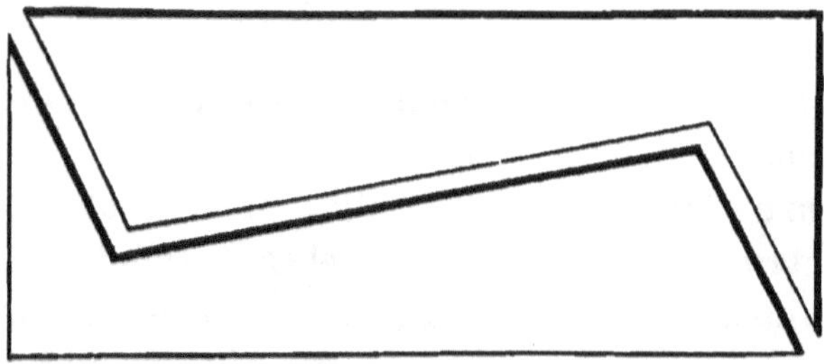

Fig. 59.—Diagram showing the mode of cutting out splints from a sheet of millboard.

inches wide above and 2 inches, or, if the limb is small, 1½ inches wide at the heel. The inner and outer strips of similar width must be cut with side-pieces for the foot, and these side-pieces stop short of the roots of the toes. For a child's thigh, the foot need not be included; it suffices for the splints to reach the small of the leg, though to prevent shortening in an adult it is usually necessary to include the whole limb. The splints are readily cut, by first marking on the sheet of millboard the required width and length of the strips, then bending the sheet over the edge of a table along these lines. The two lateral splints may be first taken from the sheet in one wide strip, after allowing for the foot-piece; the two strips are separated through a diagonal line, so that the broad end of one splint is taken from the other (see fig. 59).

When the strips are cut they should be laid on a large tea-tray, boiling water poured over them, and a minute or two later, some boiling hot thin starch; this soon soaks into and softens the millboard till it is thoroughly pliant. When somewhat softened, the edges should be thinned by peeling off little strips along them, after which some more boiling water may be poured on and allowed to soak in while the limb is prepared.

Step 2. The limb is washed and dried; a strip of diachylon plaster one inch wide is laid along the front to protect the skin when the case is being cut open after it is dry; the limb is next wrapped evenly in cotton wool, putting a scrap between each toe. This is best done by unrolling a sheet of wadding, splitting the sheet into a layer of suitable thickness, and tearing it into strips about three inches broad, which are then wound evenly round the limb as high as the splints will reach.

Step 3. The splints are next adjusted and moulded to the limb, being temporarily secured by a few ends of bandage tied round them. One assistant grasps the splints and foot at the ankle and keeps up extension, while another holds the thigh. The surgeon then proceeds to roll the bandages, first round the foot and ankle, and then up the leg, rubbing in the *warm* starch as he proceeds.   Each turn of the roller should be made as tightly as possible, for when the case dries it always grows loose by the evaporation of the water it holds. As reverses are always difficult to cut through afterwards, they should be avoided, and the bandage laid on in simple spiral or figure of 8 turns.   When the perinæum is reached, the surgeon wraps round the pelvis a broad and thick strip of cotton wool, while an assistant on each side of the patient supports his body on a folded sheet or jack-towel, and a third holds the broken limb.   The bandaging is then continued in a well-fitting spica, and ended by a few circular turns round the body.   If the splint touches the crest of the ilium it should be shortened till it clears that point, or it will gall the patient afterwards.   A fold of soft lint in addition to the cotton wool should line the splint at the perinæum, or the sharp edge of the bandage, when it is dry, will chafe there also.   When the first bandage is complete, the limb should be smeared again with starch, and a dry bandage rolled over it from below upwards, which must be similarly saturated with starch as it is laid on the limb, and when finished the whole is well covered with starch.

If the fracture is recent, and no union has taken place, a long splint should be put on outside the case, fastened to the foot and extended by a perineal band while the starch is drying, that the limb may not shorten.   With children it is best to apply the wooden splint in all cases, as they are apt to wriggle about, or sit up in bed and disarrange the case while it is in a pliant condition. If the wood splint is not used, the limb should be supported in a good position by sand-bags laid along its sides.

In three days the starch is quite dry, but the drying may be hastened by hot-water bottles or hot sand-bags laid in the bed.* The case must then be cut up along the front from bottom to top; it will often be found loose, especially where swelling had existed before.

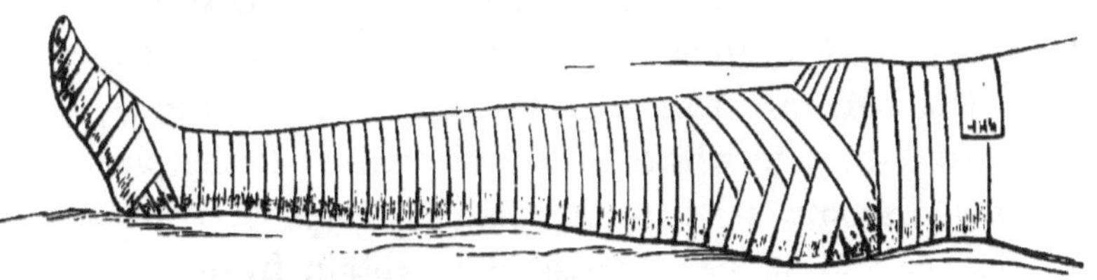

Fig. 60.—Starch Bandage.

This is best remedied by paring the overlapping edges with scissors. If any projecting part is chafed, an accident that ought not to happen, the case may be lifted from the sore part by a little more wool laid around, *not on* the part pinched. The limb being in a satisfactory position, and the case fitting properly, a roller is carried up over the whole to keep it in place while it is worn (fig. 60).

The patient need not now be confined to bed; on the contrary, the limb should be supported by a sling round his neck, while he gets about with crutches, if his leg be the part injured.

The fracture should be examined from time to time,

* Yandell finds that the following batter dries in two or three hours:—Beat the whites of one dozen eggs to a stiff froth, then stir in flour briskly to a moderately thick batter.

and at the end of three weeks some of the joints previously confined in the splint may be released by cutting off the part covering them; but if the part is a dependent one, such as the leg, it should be supported by a bandage after the splint has been removed. The limb may also be washed with soap and water, and then anointed with simple ointment, if the skin be roughened or irritated by long confinement.

In six weeks the starch splint may usually be discarded, and a roller alone worn for a few weeks longer.

Plaster of Paris Bandage. *Apparatus.* — 1. Freshly-burned white plaster of Paris. If the plaster have become stale by keeping in improperly closed vessels, and it be impossible to obtain fresh plaster, the water the plaster has absorbed from the atmosphere can be driven off by heating the powder in an ordinary oven, or before the fire, to 200° F. or 260° F., but not higher, as greater heat destroys the power of "setting." 2. Rollers, about $2\frac{1}{4}$ inches wide, of "crinoline" muslin, (*i.e.*, coarse with an open texture,) 3 yards long. 3. A roller of welsh flannel 3 inches wide and 6 yards long. 4. Basin of cold water deep enough to submerge the bandage when set on end. 5. Soft lard or spermaceti ointment and a kitchen spoon.

Step 1. The muslin rollers are best prepared by being loaded with dry powder just before they are used. To do this the roller should be gradually unrolled on a table while one person rubs in the powder, and a second rolls the loaded bandage up again loosely. While this is being done, the limb should be thoroughly washed and dried; supported, if the fracture be recent, by sand-bags. When everything is ready, a bandage should be set on end under water till all the air has bubbled out, and is then ready for use ; while the first is being applied another is put to soak, and so on.

Step 2. The surgeon carefully greases the limb wherever the plaster will reach, and rolls a welsh flannel roller round it for about 3 inches at the points where the plaster roller will cease. This protects the skin from the rough edge of the case when the appa-

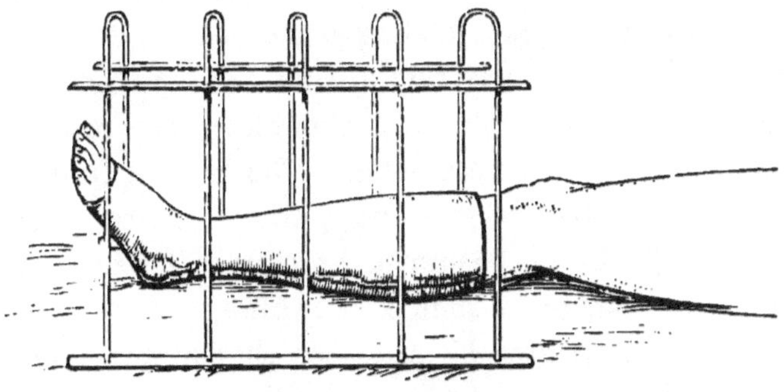

Fig. 61.—Plaster of Paris Bandage for simple fracture of the tibia, and common bed cradle.

ratus is set and hard. Indeed, if the whole of the surface to be covered with plaster be enveloped in a flannel roller, the apparatus is more comfortable to the patient and the grease may be dispensed with. When the limb is prepared the surgeon intrusts it to assistants, who will maintain reduction while he lays on the plaster rollers, rubbing the folds down closely to the limb as they are laid on. Usually two or three layers of roller give sufficient rigidity to the apparatus ; but if the limb is heavy, the case should be strengthened, by smearing over it an additional coating of plaster. This is prepared by shaking the powder into a basin of water kept constantly stirred, till it has the consistence of thin cream. The surgeon must watch that the fractured bones are kept in position till the plaster is set ; a process sufficiently advanced in five minutes for the bandage, supported by sand-bags, to be left till dry.

When the thigh or hip are to be bandaged the patient's body must be made accessible to the roller. If two stiff uprights 2 feet long be screwed to the end of a firm table and a piece of broad bandage fastened to them at their ends, a sling is formed on which the patient's buttocks may rest during the application of the bandage. His loins and shoulders may rest on a mattress or coiled-up bolster. The broken limb can be extended by weight and pulley carried over a standard (as in fig. 58) until the plaster has set. The sound leg rests on a chair meanwhile.

When the plaster is quite set the broken bone is immovable, and may be carried about without risk of displacement. In deciding what joints should be included in the bandage, the same rules obtain for this as for the starch bandage. No more joints should be rendered immovable than are necessary to obtain command of the broken bone. When the fracture is near a joint, that joint must be confined to prevent the bones from moving. When the fracture is near the middle of a bone, sufficient control can be exercised to prevent the broken ends from moving, while the joint above remains free.

To cut a Window. — If the plaster apparatus is applied over a wound, the latter should be covered with greased lint, and its position noted before the rollers are applied ; when the apparatus is set, the plaster must be dissolved around the wound by touching it with strong nitro-hydrochloric acid ; when the acid has been carried completely round, the isolated fragment of plaster may be removed, and the wound exposed.

For removal, the splint can be softened by acid along a line, and slit up with scissors, or sawn through, when the apparatus will come off in a piece.

Mr. Walker, of Peterborough, has invented a machine for passing the dry bandage through a cream of equal weights of dry plaster and mucilage (the latter prepared according to the directions given in the British Pharmacopœia). This mixture sets slowly, and gives the operator more time for applying the bandage. The machine rolls the sodden bandage as it issues from the cream on a spindle ready for use.

Common calico rollers can be made to answer the purpose tolerably well when bandages of loose texture are not at hand.

Croft's Plaster of Paris Splints for the Leg.—This plan is an improvement on the Bavarian plaster case, which has never been widely adopted in this country. It has the advantage of being applicable immediately after the accident.

Apparatus.—1. Four strips of house flannel or old blanket, in width equal to half the circumference of the limb and long enough to reach from the thigh above the knee to the toes. 2. Two $2\frac{1}{2}$ inch muslin rollers, 6 yards long. "Butter cloth," *i.e.* the gauze in which London buttermen wrap fresh butter, is the best material for the bandage. 3. A packet of freshly-burned plaster of Paris. 4. A basin of cold water.

Step 1. Wash and dry the limb as it lies on a couch supported by sand-bags.

Step 2. Take the patient's stocking and lay it out flat on a table. This serves as a guide for the shape of the flannel strips, which must be made in pairs, to fit the leg sideways, but leaving $\frac{1}{2}$ an inch of the leg in front uncovered and not overlapping behind. Thus two inside lateral splints and two outside ones are made to fit the leg and foot from just above the knee to the roots of the toes.

Step 3. Lay an inside and an outside strip on a table, outer surface uppermost, and shake plaster powder into a wash-hand basin half filled with water, stirring constantly till a thin cream is formed. Take one outside and one inside splint and soak them thoroughly in the cream, squeezing the air well out of the flannel. Next lay the soaked strips on their corresponding dry fellows.

Step 4. Reduce the fracture, and, while the broken bones are kept in position, apply the splints to the limb, the dry surface next the skin, moulding them well to the surface and irregularities of the limb. Still maintaining proper traction on the bones by means of one or two assistants, roll on the dry muslin rollers in spirals and figures of 8, evenly, not tightly, and so that no one turn is tighter than another. One thickness is enough.

Traction must be kept up till the plaster has set, which takes place in a few minutes. When set, the limb is laid in an easy position on a pillow, toes upwards. In two weeks the patient may get on his crutches. When it is desired to remove the splint for any reason, the bandage is slit in front by scissors, and the two halves of the case opened by bending at the back, where the bandage, being supple, acts as a hinge.

To ensure that the fracture shall be thoroughly reduced, and that the fragments are kept in good position till the plaster is firm enough to hold them still, the patient's muscles should be made passive by administering ether or chloroform to him.

Before the patient gets about, the surface may be gummed, or a fresh roller gummed on it, to prevent crumbling down of the somewhat brittle surface of the case.

Various stiffening materials, such as gum thickened with powdered chalk or glue, paraffin, &c., are also employed for stiffening bandages and flexible splints, after they are moulded to a limb, but none of them are as readily procured or have much advantage over starch and plaster of Paris. Silicate of soda stiffens the bandage sufficiently to need no pasteboard splints, but it requires as much time as starch to set and dry. A detailed description of the mode of using them is unnecessary.

Sand-bags are very useful, when laid along an injured limb, to prop it up on either side. For this purpose they are better than pillows, as their weight prevents their slipping from under the part they support. They should be made of macintosh cloth, be about 4 or 6 inches thick, and in length vary from 1 to 4 feet. The seams should be well closed that the sand may not escape. The macintosh should be covered with flannel, renewed from time to time.

The sand should be washed and well baked before the bags are filled, that it may not rot the cloth containing it. The bags should be only three-quarters full, or they will be too hard to adapt themselves to the limb when in use.

Sayre's Plaster Jackets. — *Apparatus.* — 1. A packet of freshly-burned plaster of Paris. 2. Knitted, elastic, woollen "skin-fitting" vest.* 3. Two-, three-, and four-inch-wide rollers of "crinoline" or cross-barred muslin, three yards long. 4. Cotton-wool. 5. Two silk pocket-handkerchiefs. 6. Sayre's triangle, with pulleys, head-collar, and arm-loops. 7. Two basins of cold water deep enough to submerge the rollers when set up on end. 8. Strips of thin tin-plate, $\frac{1}{2}$-inch wide, and punched with holes to roughen their surfaces. 9. Safety pins; needle and thread. 10. Towels. 11. Black lead or crayon pencil.

Besides the above, Welsh flannel rollers, and thick, soft felt are often useful to protect the skin overlying bony projections from being chafed under the plaster case; and an apron with breast-bib and sleeves to protect the operator's clothes.

Step 1. From 6 to 12 rollers, according to the size of the patient, are prepared, as directed on p. 92.

Step 2. The suspending triangle is erected; the basins of water, bandages, pins, and other articles are all set ready.

Step 3. The patient, stripped to the skin as low as

* The "Skin-fitting" vests are made by the American Machine Knitting Company, Oxford Street, London, and are sold in sizes by all surgical instrument makers.

the top of the thigh bones (great trochanters), pulls on the skin-fitting shirt. The upper edges are drawn towards each other as high as the top of the shoulders by tying the shoulder-straps together, and, if needed, by a few stitches carried backwards and forwards at the root of the neck. Then, beneath the shirt and next the skin, a pad of cotton-wool folded in a silk handkerchief is slipped up in front, leaving the ends of the handkerchief hanging below the shirt. This pad, termed the "dinner-pad," is used to preserve a slight concavity over the pit of the stomach and the navel after the plaster has set. In women two small pads should overlie the breasts. When the pad is put in, fasten the shirt down in front and behind with safety pins to a folded towel passed between the thighs. A towel should then be wrapped round the patient's legs to protect the dress from falling water or plaster. The level of the armpits while the arms hang down, is then marked in pencil across the back and front of the shirt.

Step 4. Arrange the patient in the suspending triangle, and let him draw *himself* up until his weight is supported by his neck and arms; the tips of his toes should rest on the ground to keep steady the body. But the patient must support himself by pulling the cord just so much as feels comfortable to him, and no more.

This being arranged, begin the bandaging. An assistant plunges into water a roller, so that as it stands on end in the basin the roller is covered with water; as soon as bubbles cease to rise up from the roller it is ready for the assistant to hand to the operator; as he does so he puts in another roller

Fig. 62.—Applying the plaster jacket during suspension.

to soak. The operator, standing in front of his patient, rapidly rolls on the bandage, beginning at the top, where the pencil mark shows the level of the armpits ; above this line the bandage must not be carried. The assistant watches the progress of the bandage behind the patient, smoothing down the turns as they are applied, that they may fit closely to the patient's body ; and wetting any part of the bandage that the water may not have reached during immersion. Bandage after bandage is thus laid on the trunk, taking care to carry the lower turns well below the hips, just avoiding the pubes and the great trochanters. When the trunk is wrapped in one layer of bandage, strips of tin-plate, long enough to reach nearly from the upper border to the lower one, are laid along each vertebral groove and below each armpit; these are sufficiently supple to bend readily to the shape of the body ; but give the case rigidity where it is likely to be strained by the weight of the patient's head and shoulders. As the strips are applied by the assistant, the operator covers them in with his roller. When about three thicknesses of bandage are applied, the jacket is thick enough for all but very heavy patients. The " dinner-pad " should be withdrawn ; the patient laid on a flat couch wheeled close to him, so that the soft plaster case may not be strained before it has hardened.* The operator, while the patient lies flat, moulds the jacket about the hips and groins by a little pressure of the hands. The patient is then covered with a light shawl or coat for

---

* A narrow table, long enough to receive the patient's full length, and swinging at its centre on a stand like a toilet looking-glass, is useful. It can be raised against the patient while he is still suspended vertically ; he may then be lowered on it to a horizontal position without risk of bending his jacket.

half an hour, or less, till the plaster has set. This interval may be employed in bending down harsh edges, turning down the upper borders of the skin-fitting shirt, and the lower ones upwards, &c. As soon as the plaster has set, the patient may be dressed and may walk about, though it is well that he take little exercise till next day, when the jacket will be thoroughly dry.

The jury mast.—When the disease in the spine is

Fig. 63.—The jury mast.

placed above the lower dorsal region, the jacket cannot be carried high enough to prevent the forward pressure at the diseased part. To obtain this, an iron support is attached to the plaster jacket while the bandage is being applied. From this support a steel rod rises behind the neck and head, and ends immediately above the vertex. Here a cross-bar, turning horizontally, projects across the head. From this cross-bar the head is slung in a jaw-bandage, and thus the weight of the head is lifted off the diseased spinal column (see fig. 63).

When it is desirable to remove the jacket it can be sawn down the front and turned one-fourth round the patient, who slips out sideways through the gap. Before it is worn again, $\frac{1}{2}$ an inch is cut off each sawn edge and replaced by a strip of leather riveted on. The leathers are perforated with eyelet-holes, that they may be drawn closely together by a lace after the jacket is put on.

Besides jackets of plaster of Paris, jackets of Cocking's poroplastic felt are also moulded to the body while hot and supple, being afterwards trimmed and fitted with lace and eyelet holes along the front. Jackets so made can be removed and replaced at pleasure. They are sold in various sizes, moulded on blocks so as to be readily adapted to the patient's figure when heated. They may be softened either in an oven or in boiling water before being fitted on. They are far more difficult to apply properly, and never fit as accurately as a plaster case.

Cradles are light arched frames of wire or cane to support the bedclothes over an injured limb. On emergency an efficient cradle can be constructed from a band-box, by knocking away part of the bottom and putting the leg through it. If used to protect a foot, a notch may be cut with strong scissors, *not* a knife, for that splits the wood.

If the cradle is stout enough, it is useful to sling a broken limb in its splint, and often great relief is thus given to the patient. Dr. Salter's Swing Cradle is specially contrived for the purpose, and is shown, fig. 41, page 66.

The Canopy Cradle is a handy arrangement, consisting of a straight wooden blade, like a large paper knife, which is placed underneath the bedclothes. The clothes and blade are then grasped from the outside by a notched block of wood suspended by an easily disposed cord and ring from the ceiling or bed frame. When this mode of suspension is not convenient, as in a low bedstead without curtains, the weight may be supported on a cord, running diagonally from the head to the foot of the bed.

Leather Splints.—For these sole leather, to be purchased at any leather dealer's, is used. In preparing them, the required length should be first noted down, then a series of transverse measurements taken at the widest and narrowest parts of the limb and over the projections of joints, &c., or a pattern may be first cut in paper and laid on the sheet of leather from which a corresponding piece is cut. The splint should always be so arranged that *its edges* do not bear on any bony point, the shin, or malleoli, for example, but either fall short of or pass beyond them. The hair side of the leather should go next the skin, as it is the smoothest and least irritating. The edges of the splint must be thinned by bevelling off the outside for about an inch all round, and no sharp corners should be left. When the leather is prepared it should be soaked, if the time can be spared, for twenty-four hours in cold water, but when wanted quickly it can be softened in a few minutes by soaking it in warm water to which a little vinegar is added—this, however, renders the leather brittle when dry, and apt to curl at the edges. When the leather is softened, a very thin even layer of cotton wadding or of lint is laid on next the skin; the splint is then moulded to the limb with the hands, and bandaged firmly; in twelve hours it will be dry and rigid. The roller is then unwound, and any parts of the splint pressing unpleasantly are marked before removal. It is then trimmed, and laid between two layers of wash-leather stitched together round the edges. The splint is now finished, and can be either fastened on by a roller or by two or more straps and buckles stitched to it.

When support is required for a joint, the splint should be fitted to the sides of the limb, where the leather has the rigidity of its width to prevent bending, instead of only that of its thickness, as at the back of the joint.

Leather Splint for the Hip.—This joint is by far the most difficult to fit. The hip splint should obtain a good grasp of its fixed point, the pelvis, and a stiff bearing on the front of the thigh where its pressure is to be exerted. There are many plans of procuring a satisfactory fit. The following is one of the best.

First cut a pattern on a sheet of paper from which to shape the leather. If possible the patient should stand while the pattern is fitting. Take a sheet of paper large enough to reach round the body, and long enough to extend from the lower angle of the scapula to the leg below the knee. Lay it against the diseased hip, with its vertical margin a little beyond the middle line in front towards the sound side, and carry the other part round the body behind, till the front is reached on the sound side. Feel for the anterior iliac spine, and mark with a pencil the point midway

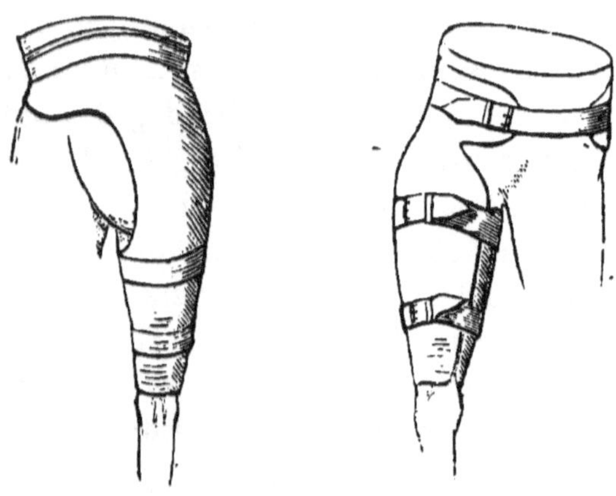

Fig. 64.—Leather Splint for the hip.

between it and the pubes; from this draw one horizontal line inwards to the border of the paper, and a second obliquely to the perinæum. Then seek for the junction of the sacrum and iliac bone behind, which corresponds pretty nearly to the point first found in front; from this mark the gluteal fold. Next carry a line vertically from the upper border of the sheet of paper to the great trochanter; and lastly, mark the level of the pelvis. Lay the sheet on a table and slit it with scissors along the lines marked, apply it a second time to the body and bend the thigh part round the thigh, making its anterior margin reach well to the inside of the limb, while the posterior part should meet it from behind. The splint should also reach downwards to the back of the knee. The paper is then trimmed down to these dimensions. The hip part is next trimmed so that it clears the buttock on the sound side and passes round till it meets the part in front. The pattern being complete, cut a piece of sole-leather to correspond, arranging that the hair or " *short* " side of the leather will lie next the skin ; bevel off the outer edge all round, and soak the leather till thoroughly soft in water : wipe it and bandage it carefully first to the trunk and next to the thigh. When it is set, superfluous and overlapping edges must be marked before removal ; lastly, the sides of the vertical notch, between the hip and trochanter, are stitched together, and the splint is covered with wash-leather.

The accompanying figure, 64, is drawn from a splint fitted by Mr. Heather Bigg on the plan just described. It has been made by the draughtsman of narrower dimensions than are advisable.

Gutta-percha may always be substituted for leather in these splints, and the same plan of fitting is used, except that the notching requisite in leather is not necessary in using gutta-percha. For the directions to use this material, see page 49.

Gooch's flexible wooden splints consist of thin laths of wood fastened side by side on stout oiled cloth or leather. They are very light, can be easily shaped, and the impermeable covering has the advantage of preventing their being soiled by discharges.

Poroplastic felt roughly shaped by the instrument makers into splints, can be obtained in sets of different sizes. They are very useful, as, when heated in an oven or in boiling water, they become quite supple, and can then be exactly moulded to the limb they are to support.

CHAPTER IV.

DISLOCATIONS.

The main obstacles in reducing dislocations are entanglement together of the displaced bones, and contraction of the muscles; the entanglement of the bones determines the direction in which extension must be made, and also that of the *counter extension,* or point at which the body is fixed to resist the traction practised on the limb; this should be exactly opposite to the direction in which the limb will be drawn. The muscles can always be relaxed by chloroform, hence it is better when they are powerful, not to use the limb as a lever to prize the head of the bone into its place. Steady extension instead is better, to disengage the bone from the parts against which it is caught, and to bring it opposite its socket, into which the hands of the surgeon guide it with less risk of laceration of the soft parts than attends forcible leverage.

Lower Jaw.—This bone is dislocated on one or both sides; when the condyle has slipped forward from the glenoid fossa, the contracted muscles prevent the bone from regaining its proper position, and cause the coronoid process to hitch against the malar bone.

Treatment.—Apparatus.—1. A towel.   2. A four-tail bandage.

The patient should be seated in a high-backed chair, resting his head against the back.   The surgeon winds the towel round both thumbs, and standing immediately in front of his patient, places a thumb on the second molar of both sides, if the dislocation be double, or on one side only, if there be single displacement (see fig. 65).   He then presses steadily downwards until the condyle

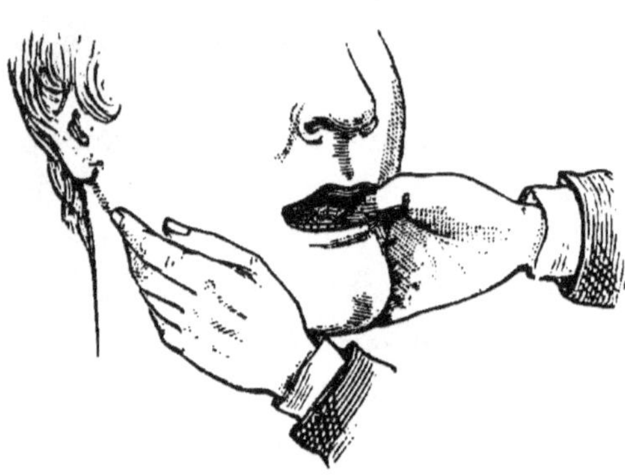

Fig. 65.—Dislocation of the jaw.

is released, when it slips back to its place.   The return of the bone may be aided by pushing up the chin with the fingers *after* the ramus of the jaw has been lowered.

When the jaw is replaced, a four-tail bandage or split handkerchief should be tied over the nucha and vertex of the head, to keep the jaw closed (see fig. 23, page 33).   Biting or chewing should not be attempted for ten days or a fortnight.   The patient should be warned also that when the jaw has been once dislocated it readily slips out of place again ; he must thenceforth avoid gaping or opening the jaw very widely.

The Clavicle is rarely dislocated, nevertheless both the inner and the outer end may be displaced.   The inner end may be driven upwards, or behind the breast bone or in front of it. The outer end may slip on to the acromion or beneath it. The signs are obvious —the end of the bone is felt in its new position. In many cases the dislocation is only a minor accompaniment of very severe crushing injury to the chest, and reduction of the dislocation is not advisable. The reduction of all varieties is very similar in its method.

Apparatus.— 1. Roller, $2\frac{1}{4}$ inches wide. 2. A piece of old blanket.

The blanket should be torn into strips about 12 inches by 18 inches, and folded thrice, thus making a long soft pad to line the axilla, one for each armpit. The patient is next seated on a stool ; an assistant standing behind, draws back the shoulders while he presses on the spine with his knee. The dislocation being reduced, the surgeon fixes the bone by a figure of 8 carried round the shoulders and across the back. The forearm is then bent and fastened to the body by a few turns of the roller round it and the chest. This prevents the pectorals from acting on the bone. The apparatus may be laid aside at the end of two weeks, but the arm must be fixed to the trunk for another fortnight.

This bone is often difficult to keep in place after dislocation, and even the most accurately fitted apparatus sometimes fails to effect its object, hence many varieties of collar and yoke have been devised by different surgeons to accomplish this purpose.

The Shoulder is dislocated in three directions, downwards, inwards, and backwards. These have subordinate varieties, the distinguishing signs of which depend chiefly on the direction of the greatest displacement.

Signs of dislocation into the *axilla*. When the bone is displaced below the glenoid fossa, the acromion is prominent ; underneath it, the surgeon feels a hollow instead of the head of the humerus, which the finger detects in the axilla. Movement of the shoulder is very limited and painful. If the elbow is rotated while the finger is in the armpit, the head will be found to move with the rest of the bone.

If the head of the bone is carried more *inwards* towards the ribs, it can be seen and felt near the clavicle ; the hollow is again readily detected below the acromion, while the axis of the arm is .altered, being directed inside of its proper position.

When the bone is carried *backwards* (a most rare accident) the head is plainly felt on the scapula below the spine.

For the reduction of these dislocations several plans are employed. When recent, the two first displacements can generally be restored without chloroform, but if the patient is muscular, it often saves time and pain to produce anæsthesia before attempting to replace the bone.

By the foot in the axilla (fig. 66).—The patient lies flat on a couch ; the surgeon, pulling off his boot from the left.foot if he has to reduce a left dislocation, and from the right foot, if that be the side of injury, seats himself on the couch facing the patient. Putting his unbooted foot into the armpit, he grasps the forearm with both hands and pulls steadily downwards. When the head of the bone is disengaged, the muscles draw it

into the socket, and the movements of the limb become
at once easy and natural.   The arm must then be fixed

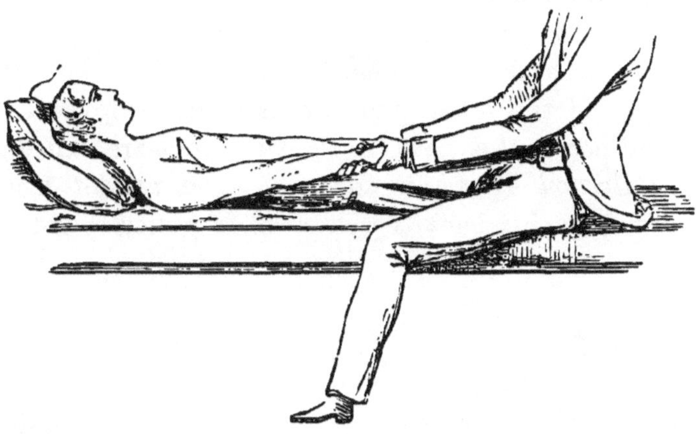

Fig. 66.—Reducing a dislocated shoulder by the foot in the armpit.

to the side by a roller for a fortnight, and the shoulder
is wetted with an evaporating lotion to allay the pain
and inflammation resulting from the laceration of the
soft parts.   Should the sur-
geon's strength be insufficient
for the requisite extension, a
jack towel may be attached
in a *clove-hitch* round the arm
above the elbow and held by
an assistant, who, standing
behind the surgeon, draws
steadily in the same direction.

To make a clove-hitch.—Grasp
the towel in the left hand, the
little finger being downwards,

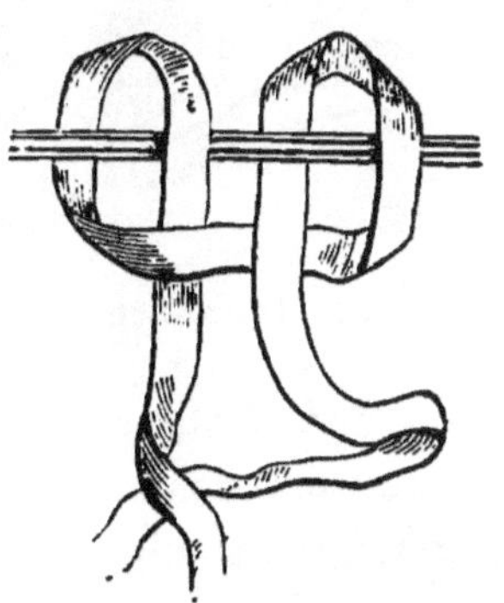

Fig. 67.—The Clove-hitch knot

then turning the right thumb down till its little finger

is upmost, seize the towel below the left hand ; if the wrists are then rotated till the thumbs come together, the towel will be drawn into two loops, of which the ends cross, on opposite sides, the connecting part between the loops (see fig. 67). If one hand hold the loops and the other pull the ends, the loops will be found not to clip, however tight the ends are pulled.

Reduction by simple extension.—The patient again lies flat on his back, a jack-towel is passed round his

Fig. 68.—Dislocation of the shoulder reduced by simple extension

body and fastened behind the opposite shoulder for counter-extension, while a second towel or skein of worsted is attached to the arm above the elbow by a clove-hitch and entrusted to two or three assistants, who are desired to pull quietly and steadily directly away from the patient's body. The surgeon meanwhile watches the process of the extension, altering its direction as he finds the head more or less engaged against the scapula, and finally with his hand thrusts the head into its socket. Sometimes there is much difficulty in getting the head back to the glenoid fossa, even when the humerus is completely disengaged from the scapula ; this difficulty is often overcome if an assistant rotates the humerus to and fro while the extension at the elbow and the pressure on the head of the humerus are steadily maintained. When the limb is replaced, it is fixed to the side as before directed.

If the dislocation has existed more than a few hours relaxation of the muscles by chloroform and steady extension of the limb directly away from the body, are more sure of success than pulling with the heel in the axilla, because greater power can be exerted more steadily than is possible by the latter mode.

The Elbow.—The signs of dislocation at this joint are tolerably evident, but there is often co-existent fracture of the olecranon or coronoid processes. Separation of the articulating surfaces of the humerus from the shaft is sometimes mistaken for dislocation of the forearm backwards.

In dislocation of *both bones backwards* the olecranon is very plainly felt behind the condyles of the humerus ; the sigmoid notch is generally to be made out, and the forearm is held nearly at a right angle. The altered relation of the olecranon to the condyles suffices to distinguish dislocation from fracture of the humerus at its lower end, where the olecranon also goes backwards, *but the condyles go with it*. The little mobility of the joint distinguishes it from separation of the lower articular surfaces of the humerus from the shaft, an accident, moreover, only met with in children.

Other distinctions of dislocation from fracture are, limited movement, difficulty of restoring the bones to their natural position, absence of crepitus ; and lastly, the peculiar form of the articular surfaces can some-times be made out.

In reducing the *backward dislocations* the patient sits on a chair on which the surgeon rests his foot, pressing his knee against the forearm at the elbow for a fulcrum ; then, grasping the wrist with one hand, and steadying the arm with the other, he flexes the elbow to dislodge the coronoid process from the fossa at the back of the humerus ; when this is done, the articulating surfaces slip into place. If this plan fails to reduce the dislocation, extension of the bones of the forearm in the axis of the humerus with the elbow bent, must be employed either by means of pulleys, or of jack-towels and the muscular strength of several assistants.

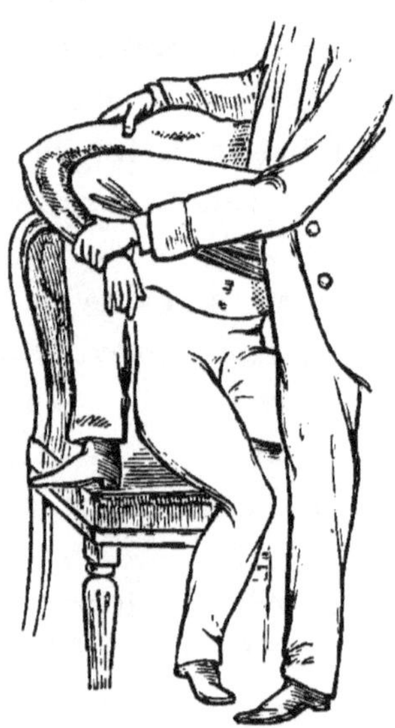

Fig. 69.—Reducing dislocation of the elbow round the knee.

The body should be fixed by a jack-towel carried under the armpit of the injured side, and over the shoulder of the sound side. A wetted bandage is rolled round the forearm, and a second towel is attached by a clove-hitch (see fig. 67, p. 113) to it just below the elbow. Extension is then made in the axis of the humerus, while the elbow is worked gently to and fro, until the surgeon's hand can push the bones into their places.

When the *radius only* is displaced, the manipulations just described may be exercised. If extension, not leverage, be used, the elbow must be straightened and the pull applied to the wrist.

In all dislocations of the elbow, when the bones are returned, the limb should be bent to a right angle and put on a lateral angular splint for a week or ten days, after which time it should be worn in a sling for a fortnight longer.

Dislocations of the wrist are extremely rare. They have generally been reduced by simple extension.

The Thumb and Fingers.—When the first phalanx is dislocated from the head of the metacarpal bone, it is sometimes very difficult of reduction. The most

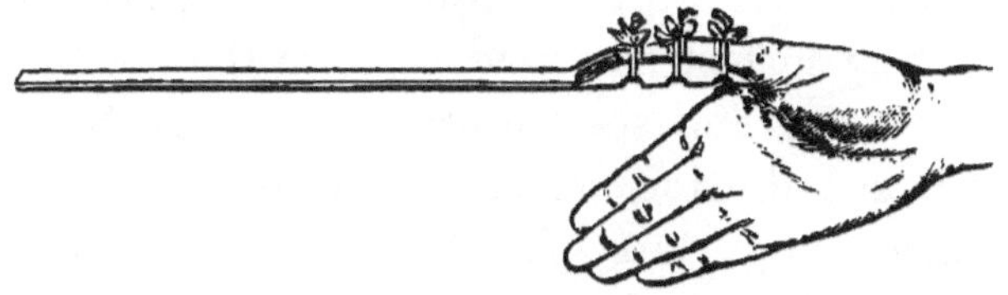

Fig. 70.—Handle for obtaining control of the thumb in dislocation.

effectual mode is steady extension, which is procured by fastening the thumb to a piece of wood, which serves as a handle to give command of the phalanx, and is contrived in the following way. The thumb is first bandaged with a narrow wetted roller over the two phalanges, and a thick layer of cotton wool is rolled round it. A piece of stiff wood, 1 inch wide,

½ inch thick, and 12 long, is perforated at one end with three pairs of holes ½ inch distant from each other and from the end ; through these, three stout tapes, ½ inch wide and 2 feet long, are threaded, leaving three loops on one side of the piece of wood (fig. 70). The wood is then applied to the palmar aspect of the phalanges, the loops passed over the thumb, their ends drawn tight, and tied, not in a bow as the figure represents, but in a knot wound round the end of the stick.   Thus attached, the stick becomes a good handle for extending the digit, and also a long lever for altering the direction of the phalanx if desired. Langenbeck of Berlin employed a pair of forceps to seize the thumb, instead of the wooden handle just described.   Leverage may be used if extension in the axis of the metacarpal bone fail. Should the phalanx have slipped on to the back of the head of the metacarpal bone the phalanx may be tilted back-wards further and further until the hinder edge of its articulating surface be prized round to the hinder edge of the articulating surface of the head of the meta-carpal bone ; if extension be now made suddenly, the bone often slips into place. If the phalanx have passed in front of the head of the metacarpal bone, towards the palm, the prizing must be carried in the opposite direction, over-flexion.   But with the greatest care and perseverance it is sometimes impossible to replace the bone unless the constricting bands be divided with a tenotome.

Hip-joint.—There are three chief directions in which the hip is dislocated. First, *backwards* on to the dorsum ilii, or further on to the sciatic notch.   In this dislocation the limb is shortened, moved with difficulty, drawn inwards over the other, and its great toe overlies some part of the back of the other foot. The hip itself is altered; the great trochanter being nearer to the crista ilii, and more prominent than on the uninjured side, and the head is often plainly felt in its new position. Resistance to extension of the limb, limited movement of the hip, with rotation inwards, are the distinguishing points between this dislocation and fracture at the neck of the femur.

*Treatment.—Apparatus.—*A complete apparatus for this purpose is contrived and sold by instrument makers, but a sufficiently serviceable one can be extemporised when the former is not at hand; it consists of :—1. A rope running in two pulley blocks. 2. Three jack-towels or skeins of worsted. 3. Two stout hooks to screw into the wall, or to some firm object, to obtain fixed attachment. 4. A wetted roller 3 inches wide.

The complete apparatus is as follows :—

*Apparatus.—*1. A set of multiplying pulleys. 2. A leathern padded girth, 2 inches wide and 3 feet long, having at each end an iron ring. 3. A stout leathern belt about 6 inches broad, furnished with buckles, straps, and rings to fasten on to the thigh above the knee ; a rope is run through the rings to connect the hook of the pulleys with the thigh. 4. Two strong iron hooks to screw into the wall, for fixing the apparatus. 5. Half-a-dozen yards of stout cord. 6. A hook, fitted with a buckle and strap, and so hinged that, by turning a pin, the hook flies open and at once slips off. If this be interposed between the pulleys and the belt fixed on the thigh, the limb may be instantaneously released when desired.

Step 1. The patient is laid on a flat couch, and put under the influence of chloroform. When he is narcotised, he is laid on his sound side and a jack-towel, or, if it be at hand, the pelvic girdle, is carried across the perinæum; it is arranged to bear on the tuber ischii behind and on the pubes in

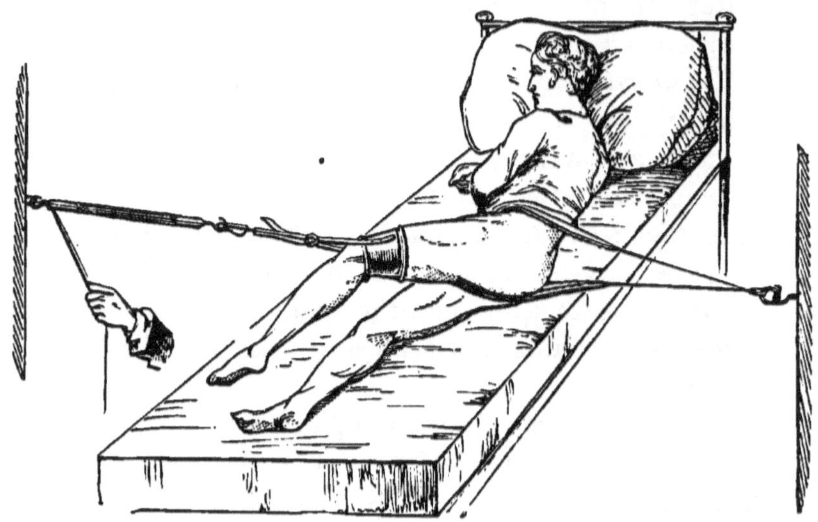

Fig. 71.—Dislocation of the dorsum ilii.

front, with its ends attached to one of the hooks screwed into the wall behind, and about six inches below, the level of the patient. This towel should be put slightly on the stretch, that the pelvis may be kept in the position first assigned to it when the pulleys begin to draw. A wet roller is put on the lower third of the thigh, and a jack-towel in a clove hitch, or, as in the drawing, a padded leathern band, slipped up the leg to the bandage. Another jack-towel is then doubled and passed up the limb to the perinæum. The dislocated thigh is bent across the middle of the sound thigh and the clove-hitch connected by the disengaging hook with the pulleys, which are attached to a hook fixed a little above the level of the patient, on a line carried from the hip across the junction of the middle and lower thirds of the uninjured thigh (see fig. 71).

Step 2. The surgeon being ready, an assistant draws on the pulley cord, getting gradual extension of the limb as required by the surgeon, who, keeping his hands on the hip and great trochanter, watches the progress of the head of the bone towards the acetabulum.

Step 3. When the bone has reached the edge of the acetabulum, a second assistant slips the doubled jack-towel over his shoulders, and by raising his body, lifts the femur away from the brim of the acetabulum, while the surgeon, grasping the foot and knee, makes a few movements of rotation backwards and forwards to ease the head into its socket.

When reduction has been effected, the limb should be put in a long splint or starch bandage for four weeks, and the patient not allowed to exercise the limb violently for a month after he walks again.

Reduction by manipulation.—When the patient is not very muscular, and the dislocation recent, the bone can often be speedily returned by movements of flexion and rotation.

The patient is put fully under chloroform and laid on a mattress on the floor. The surgeon, standing on the injured side, grasps the small of the leg by the front, and passes his other hand behind the knee, bending the leg till it is at a right angle with the thigh : the thigh is bent over the belly. Next, he adducts the thigh a little and rotates it inwards by turning out the foot, to disengage the head from behind the socket. The third step is to carry the knee outwards. When the thigh is perpendicular, the abduction is changed for circumduction and rotation outwards. During these latter movements the limb is jerked towards the ceiling- by the hand placed behind the knee. Lastly the limb is straightened as the head drops into the socket.

The pelvis may be steadied by the surgeon's foot, divested of its boot, placed on the iliac spine.

Dislocation downwards into the thyroid foramen. The limb is usually lengthened, capable of little motion ; the knee is bent ; the toe points forwards, and away from the other foot. Here the reduction is best managed by extension ; the apparatus required being the same as that employed in dislocation backwards, but it is differently arranged.

Step 1. The patient lies on his back, the pelvic girth, or towel, is carried round the pelvis and fastened to the wall on a level with his body, opposite the uninjured side. A jack-towel is put round the upper part of the dislocated thigh, and attached to the pulleys outside, which are fastened to the wall opposite (see fig. 72).

Step 2. Extension is then made by an assistant: the surgeon grasps the leg above the ankle, and rotating the limb inwards and outwards, but without lifting it from the bed, guides the head into the acetabulum.

Here, as after dislocation backwards, a long splint should be worn on the limb for three weeks before the patient is allowed to move about at all.

Reduction by manipulation.—The patient being prepared as above directed (p. 121) the surgeon flexes

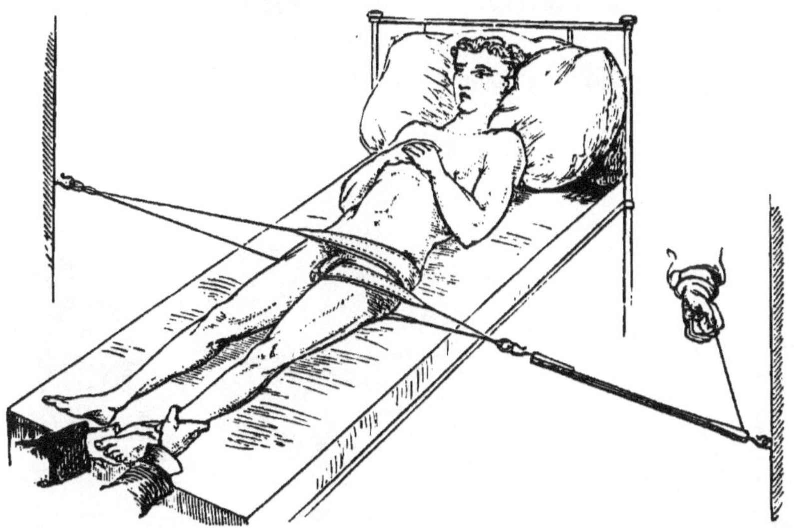

Fig. 72.—Dislocation downwards.

the limb towards the perpendicular and abducts it slightly; he then rotates the thigh strongly inwards, at the same time adducting it; and, carrying the knee towards the floor, finally straightens the limb.

Dislocation on to the Pubes.—The limb is easily moved at the hip, shortened, rotated outwards, and the head of the bone is felt in the groin.

The same apparatus is used in this as in the dislocation on the dorsum ilii. It is applied as follows :—

Step 1. The patient lies on his back (fig. 73), with his legs separated. The pelvic band is passed over the perinæum and pubes, and attached above the patient, in a line passing from the pelvis a little to his sound side. A doubled jack-towel is slipped up the limb to the perinæum ; the pulleys are fastened to the thigh above the knee and fixed, in the manner directed on page 120, to the wall below and external to the injured side of the body.

Step 2. Extension is then steadily made, while the

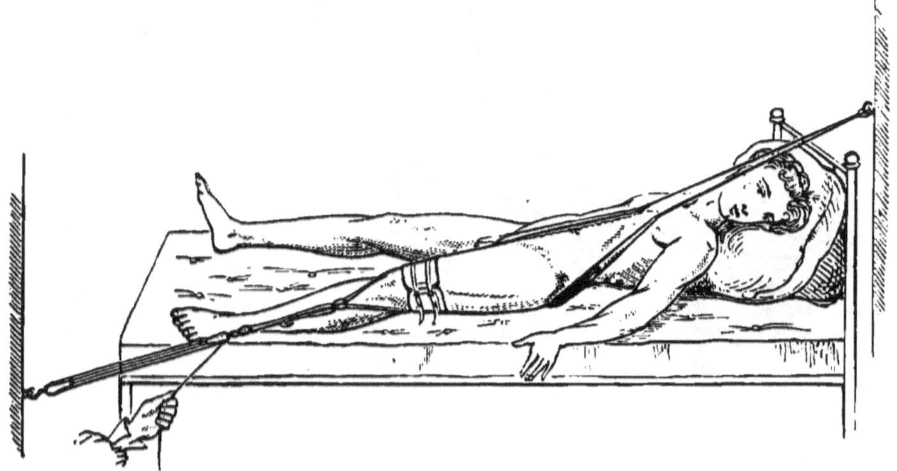

Fig. 73.—Dislocation on to the pubes.

surgeon watches the head getting free from the pubes, over the edge of which a second assistant, slipping his neck through the doubled towel, raises the bone a little outwards. The surgeon in the meantime encourages the bone by rotation to enter the socket.

Reduction by manipulation.—The patient being placed

on his back, the surgeon grasps the leg with one hand and the upper part of the thigh with the other; he then draws the limb downwards, at the same time flexing it gradually on the abdomen as far as possible; he then rotates the thigh inwards; and directing the head of the bone by its shaft, gradually rocks it downwards into its place.

A splint is necessary here also after reduction.

The Knee.—These dislocations are rarely complete. The lateral ones are often easily reduced by flexing the thigh on the belly, straightening the leg, and rotating it a little from side to side.

Another Plan.—*Apparatus.*—Two jack-towels. This is more useful when the tibia is carried backwards. Lay the patient on his back, and slip a jack-towel in a clove-hitch up the leg to the thigh, and another round the small of the leg; the thigh is bent and retained in an upright position by an assistant holding the jack-towel at the ham, while a second pulls on the one at the ankle and so disengages the bones from each other, when the surgeon readily slips them into place.

After reduction is accomplished, the limb should be fixed in a leathern back splint until the inflammation subsides.

Dislocation of the Patella.—The displacement of this bone on to the outer or inner condyle is generally easily reduced if the knee be straightened and the vasti muscles relaxed by bending the thigh on the belly. When the patella is turned on its own axis, the side, not the under surface, is locked against the condyle, and reduction is sometimes extremely difficult or impossible.

The same movements must be adopted as for simple lateral displacement, and the surgeon must endeavour to release the bone by pressing its upper edge downwards with his thumbs.

After their reduction, all dislocations about the knee-joint must be treated by rest, straight splints, and evaporating lotions.

The Foot is very rarely dislocated from the leg without fracture of the malleoli. Its reduction requires simple extension of the foot on the leg, with the knee bent ; the surgeon grasps the heel in one hand, the foot in the other, while an assistant fixes the thigh in the half-bent position. The foot is first drawn downwards to disengage it from · the tibia, and then directed into its place.

After reduction the limb should be put in a McIntyre's splint, in the way described for fracture of the tibia near the ankle-joint.

Reduction is often impossible, and the displaced bone must be excised.

Scarpa's Shoes are instruments for restoring deformed feet to their natural shape. The shoe (fig. 74) consists of a flat metal sole broader and longer than the foot, furnished with a rest for the heel. A rod, attached

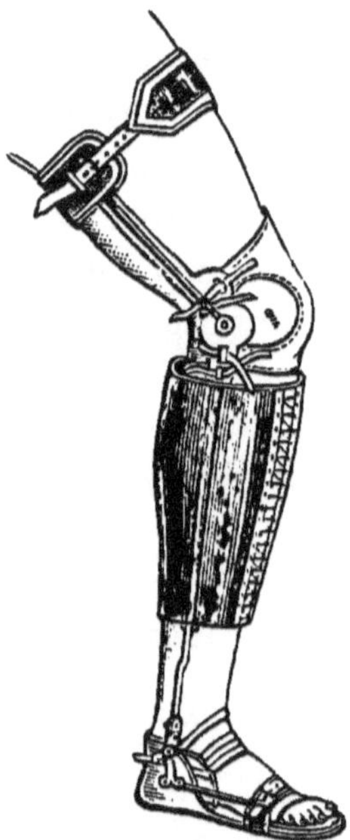

Fig. 74.—Shoe for reducing talipes equinus.

to the side of the sole beneath the ankle, reaches up the limb, to which it is secured by one broad band and buckle below, and by a second above the knee, opposite which joint the iron stem moves on a free joint backwards and forwards. In the drawing the band for the leg has been replaced by a lacing boot-leg to obtain greater steadiness of the boot on the leg. Opposite the malleoli are set the centres of movement required for the restoration of the deformity ; they are moved by a key. The foot is fastened to the sole by straps across the instep and ankle ; the toes are restrained by a strap passing round them and fixed to a horizontal toe-bar by the side of the foot.

The foot must not be very tightly braced into the shoe ; in children, if the straps are drawn tight the skin almost invariably inflames, and even sloughs where it is compressed. Before the instrument is applied, the limb should be bandaged with a soft cotton, or domett's flannel roller. The foot is first fixed to the sole or shoe, and then the leg to the rod. Traction is increased gradually with frequent small alterations, as the foot yields to the tension and regains its natural position.

Thomas's Splints.— There are two kinds in common use, the " hip splint " and the " knee splint." The former is used for disease of the hip-joint and for fracture of the neck of the femur; the latter for disease of the knee and fracture of the shaft of the femur.

The Hip Splint (see fig. 75) consists of a flat band of malleable iron, $\frac{5}{8}$ to $\frac{3}{4}$ inch broad, $\frac{3}{16}$ inch thick for light children ; $\frac{3}{4}$ to 1 inch broad and $\frac{4}{16}$ inch thick for heavy lads and most adults ; (for exceptionally heavy adults the band may be required $\frac{6}{16}$ inch thick and 1¼ or even 1½ inch wide). In all cases it is indispensable that the band should be so rigid that it will not yield to any effort of the patient, though it can be bent where necessary by the surgeon. This band reaches from the lower angle of the scapula to

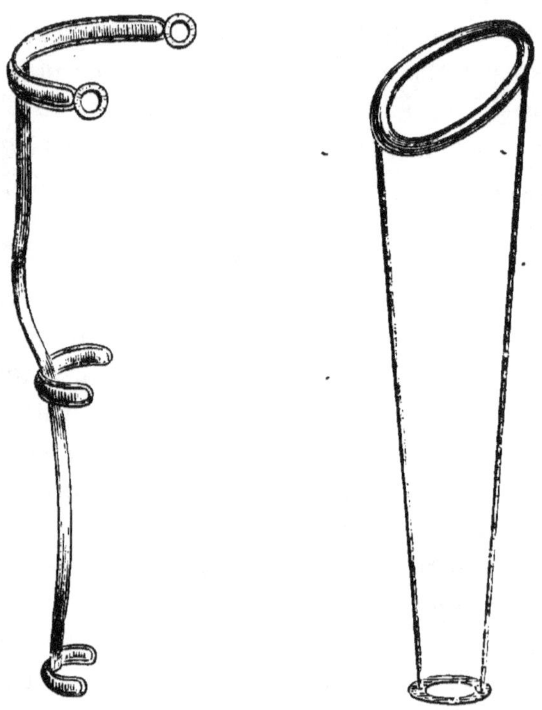

Fig. 75.—Hip Splint.   Fig. 76.—Knee Splint.

the middle of the leg. It passes behind the thorax and loin, crossing the buttock just behind the neck of the femur, and is continued downwards behind the centre of the limb. At the upper end a horizontal flat bar about 1 inch wide encircles the chest, as far forwards as the middle line on the side of the diseased hip and as far as the nipple on the other side. This chest band should never be less than ⅛ inch thick, and that will do for all sizes, as it must be rigid enough to keep its shape when all is fastened up. The ends of the horizontal bar are connected by a strap and buckle in the older splints, but they are now preferably finished off with an iron ring, ¾ to 1 inch wide inside, welded to the bar, whipped with thread and varnished. Opposite the upper third of the thigh, and at the lower end of the splint, horizontal bars, ½ to ¾ inch wide and ⅛ inch thick, are placed. They embrace about two-thirds of the limb. The whole instrument is covered with basil leather, between which and the iron no more padding is required than a single strip of thin felt or thick coarse flannel. A pair of braces over the shoulders holds up the instrument when the patient walks about, and prevents its slipping down in bed. This bracing is *always* essential, and is best effected by a bandage of swan's-down calico, 3 or 4 inches wide, with which the chest band is also closed by passing it through the rings, when they exist.

The instrument is bent, when necessary, by means of strong iron wrenches. When the hip is much flexed, the splint is bent to fit round the buttock without pressing specially on any part. The adduction, abduction, inversion or eversion of the limb, as a rule, right themselves as the stiffness disappears; but the comfortable apposition of the three cross-bars is effected by wrenching the splint to the necessary position. The object of this precise adaptation is to

K

relieve the strain caused by the weight of the limb on the injured part. The splint is wrenched or bent to follow the improvement in the position of the limb as often as the relaxation of the joint permits such

Fig. 77.—Showing the hip-splint, with patten and long crutches.

change. But this is usually only required in the longitudinal stem for reducing deformity in cases of excessive flexion of old standing. In all cases where the flexion is slight, or the disease quite recent, the splint should be put on nearly straight. If an acute flexion with great tenderness then be present, insert a chaff pillow between the *bent* knee and the splint, and bandage all together. This will accommodate the flexion of the hip, but can be diminished almost daily, and within from three to ten days dispensed with altogether, without altering the shape of the splint. The joint is thus steadied at first in the flexed position, and the previous pain considerably mitigated *at once;* while by gradually inserting a less and less thickness of pillow under the knee, the straight position, the only sound curative one, is quickly reached. The return of the limb to a natural position is assisted by making the horizontal bars press the trunk and limb slightly against the sides to which they are drawn; this pressure must, however, not be great enough to be irksome. When all pain and tenderness have ceased and the hip-joint is nearly straight, the patient may walk on crutches, the limb being raised clear of the ground by attaching an iron patten, 3 or 4 inches high, to the boot of the sound limb in children; and in adults by the similar use of a wooden thick sole and heel (2 or 3 inches suffices) (see fig. 77). But until this stage is reached, the patient should retain the horizontal posture. The treatment by this splint requires several months or even years, but cure, either by recovery of normal movement or by ankylosis, is surely obtained.

The Knee Splint (fig. 76) carries out the same principles of thoroughly supporting the limb and permitting, when well applied, the attainment of almost perfect linear immobility in the horizontal posture. It also prevents the foot from touching the ground, so that the weight of the body is transmitted from the tuber ischii to the ground through the iron rods and not through the knee. Two iron rods, $\frac{3}{16}$ to $\frac{3}{8}$ inch thick, are joined 4 or 6 inches below the foot by an iron ring of 2 or 3 inches diameter. The upper ends are attached to an oval padded ring, shaped, when the patient is able to walk on crutches, to surround and fit the thigh without clipping at any point. At the inner side, where the ring is most thickly padded, it bears against the tuber ischii; at the outer side it arches above the tip of the great trochanter. In front, the ring lies against the fold of the groin, but without compressing the parts. Thus the upper ring encircles the hip obliquely. By the prolongation of the rods beyond the foot, a point of attachment is provided for stirrup extension of the tibia if that is desired, and also the foot is prevented from touching

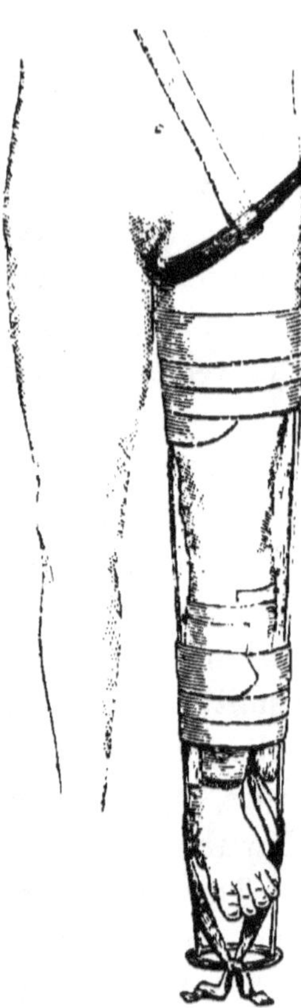

Fig. 78. - The knee-splint applied.

the ground when the cure is sufficiently advanced to permit the patient to walk about. A patten must then be applied to the boot of the sound limb to elongate that side equally with the diseased one.

The limb is supported in the splint with flannel or swan's-down rollers carried round the thigh and the leg. It may be variously applied so as to favour reduction of the tibia to its normal position, when drawn backwards towards the ham or rotated outwards. The splint may be adjusted to the limb in several ways, but in all the same principle obtains, viz., to transmit the weight of the trunk to the ground through the splint and not through the knee-joint. That part is kept at rest, though not so much confined as to prevent slow and usually spontaneous return of the tibia to its proper position on the femur.

Casting in Plaster of Paris.—It is often convenient, when ordering an apparatus for deformity, to send the instrument-maker a cast of the deformed part. This is readily made in the following way :—

Apparatus.—1. Two packets of freshly-burned plaster of Paris. 2. Some pasteboard, an old bandbox, or several newspapers. 3. Olive oil. 4. A basin of cold water.

Step 1. The part to be modelled should be laid in an easy position, thoroughly oiled, and a shell or trough of pasteboard roughly built round it to contain the plaster till it sets.

Step 2. The plaster is then prepared by shaking the powder into cold water, till a thick cream without lumps is formed ; this smoothness is secured by constantly stirring the water as the plaster is shaken in. The cream is then poured into the trough, little by little, that it may make its way into the inequalities and recesses, until the limb is half immersed, leaving the projecting parts, such as joints, half exposed, so that the halves of the mould may separate opposite them. This first instalment is then allowed to set, and a fresh supply of plaster is prepared.

Step 3. The surface of the hardened mould is oiled, that the fresh cream may not stick to it, and the whole of the limb is then covered by pouring the cream on a second time. Plenty of plaster should be laid over the projecting parts that the mould may be strong enough for use. It should be ¾ inch thick everywhere, and 1 inch thick along the sides. When the second half is set, the trough or shell is cleared away, and the two halves of the mould removed from the limb separately.

or casting, the mould is well oiled inside and filled

with cream, which sets into the cast required. While the plaster is liquid the mould should be well shaken, that the air bubbles may be all driven from the surface of the cast.

In many cases a mould of the limb taken in gutta percha gives a cast sufficiently accurate for the instrument maker. It is a more expeditious and cleaner method than that just detailed. The mode of taking gutta percha moulds has been described at p. 49.

CHAPTER V.

— ♦ —

MISCELLANEOUS.

Ophthalmoscopy practically comprises two procedures—(1) Oblique Illumination, and (2) Examination with an Ophthalmoscope. For the former, an ordinary bi-convex lens of about $2\frac{1}{2}$ inches focus is needed; for the latter, a perforated glass or metal concave mirror of about 10 inches focus, fitted with ocular lenses, as in the simple ophthalmoscope of Liebreich, or the more elaborate instruments of Loring, De Wecker, Landolt, Gowers, and others.

Oblique illumination consists in concentrating rays of light, by means of a convex lens, obliquely upon the cornea, thereby discovering the state of the cornea, anterior chamber, iris, pupil, and crystalline lens with its capsule.

In Ophthalmoscopy proper there are two modes of procedure—(1) the *direct*, or examination of an erect " virtual " image of the interior of the eye, and (2) the *indirect*, or examination of an inverted " real " image.

By the *direct* mode we may discover the condition of the cornea, and other refractive media, and of the fundus oculi. In this mode light is thrown into the eye with the mirror from the flame of a lamp, or, preferably, from an Argand gas-burner, which should be placed alittle behind, and a few inches away from the corresponding ear. If the left eye be under examination, the observer should use his left; if the right, then he should use his right eye, so as to avoid actual contact with the patient's face. It is necessary to adjust the distance according to the situation of the particular part of the eye to be examined. To examine the cornea the observer must stand from ten to eighteen inches off, but to search the posterior structures, he must gradually approach the eye, until for the fundus itself he must come close to the patient's brow. For the anterior planes of the eye, and for the fundus of an emmetropic,* or of a slightly hypermetropic eye, no ocular lens is required behind the mirror if the observer be himself emmetropic; but if he or the patient be highly hypermetropic or myopic, a correcting lens must be used.

The precise strength of the lens will vary with the refractive conditions of the observing and the observed eyes. Speaking generally, the weakest concave lens (to correct myopia), or the strongest convex lens (to correct hypermetropia), with which the details are clearly seen, is the appropriate lens. All the media having been examined—first, with the eye at rest, and then in rotation—the fundus must be considered— namely, the size, shape, outline, margins, level, and colour of the optic disk; the relative size, distribution

---

* *Emmetropic* means that the focus of the lens-system is at the retina; *hypermetropic*, that, the eyeball being too short, the focus lies behind the retina; *myopic*, that, the eyeball being too long, the focus is placed in front of the retina.

and appearance of the central blood-vessels ; the condition of the peripheral portions of the retina and choroid ; and, lastly, the region of the yellow spot. This may be seen if the patient gazes at the image of the flame in the ophthalmoscopic mirror.

In the *indirect* examination, an objective lens is employed in addition to the mirror. The observer stands about twelve to eighteen inches from the patient, and places before the latter's eye a bi-convex lens of $2\frac{1}{2}$ inches focal length, so as to form an inverted aërial image of the disk between the objective lens and the mirror. To bring the disk into view, the patient should direct his gaze slightly upwards and inwards, while the observer himself looks straight before him. Here, too, the other portions of the fundus must be carefully scrutinized. The advantage of the *indirect* method is, that it gives a more extensive view than the *direct*, and thereby affords a better criterion of the relative size, shape, and position of objects.

Fig. 79.—Eye Douche.

The **Eye Douche** is a small elastic bottle fitted with a nozzle and flexible tube, ending in a rose, through which, by means of a valve, the water is drawn from a vessel and driven in a fine spray over the eye held open to receive it (see fig. 79). The syringe in fig. 79 is very useful for a variety of purposes.

Syringing the Ears is best performed by a syringe having a long nozzle to direct the current of soap and water down the meatus to the wax. Figure 80 consists of a syringe with double opening and air

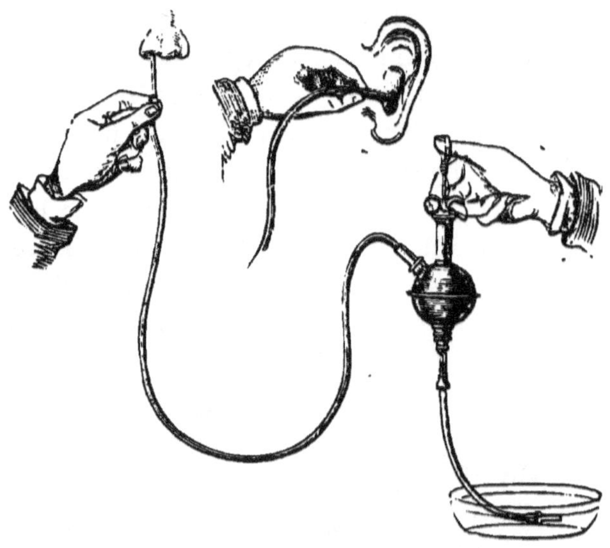

Fig. 80.—Syringe for sending a continuous current into the nose or ear, &c.

chamber; when in action it supplies a continuous gentle current, which breaks up the concretions more speedily and with less discomfort to the patient than the intermitting jet of a common syringe; but an important part of the apparatus is the long slender nozzle to direct the stream well into the meatus. The instrument-makers supply a little spout or shoot to hang under the ear, to turn off the water into a basin clear of the neck. If time permit, the patient should keep the ear charged with olive oil for a few days before syringing, that the wax may be softened. After the wax is removed, the irritation of the canal is best allayed by a little glycerine or olive oil put into the meatus, and covered by a pledget of cotton wool, large enough to fill the concha and too large to enter the passage, where it may be lost sight of.

For the Nasal Douche, a convenient form consists of $1\frac{1}{2}$ yards of ordinary india-rubber tubing $\frac{1}{2}$ inch in diameter, to one end of which is attached a perforated metal conical weight, and to the other an ordinary broad-shouldered nozzle. The weighted end rests at the bottom of a vessel containing the injection, placed on a level above the patient's head. The tube is then made to act as a syphon, delivering the fluid through the nozzle in a continuous stream. The patient should hold the head forward and the mouth open over a basin, as the soft palate thus closes the pharynx, when, if the nozzle be fitted closely into the nostril, the fluid passes from one meatus to the other in a forcible stream, which cleanses the nares and upper part of the pharynx thoroughly and issues on the face from the open nostril.

Only weak solutions of astringents or of common salt having a density equal to that of the blood, should be employed to wash out the nasal passages.

Ice-Cold Injection.—In obstinate epistaxis the nares are sometimes plugged, but before proceeding to this painful mode of treatment, a simpler plan should first be tried; namely, the *injection of ice-cold water* into the nostril along which the blood flows.

The stream should be directed upwards that the water may first dislodge the clots entangled in the meatuses, and then flow over the bleeding surface. This is best done by employing tbe nasal douche above described (see fig. 80), or a clyster syringe, one tube of which lies in a vessel of ice-cold slightly salted water (containing solution of gallic acid or other styptic if desired, though cold water alone usually suffices), the other tube, having a long narrow nozzle, is passed up the nostril and directed upwards among the spongy bones. With this apparatus the water is injected steadily for half an hour, before being abandoned as unsuccessful. The patient is kept still, sitting upright in a cool room. If these means fail to check the flow of blood, the nares may then be plugged.

Plugging the Nares.

Apparatus.—1. A flexible catheter, No. 7, or Belloc's sound. 2. Whipcord. 3. Lint. 4. Scissors.

Step 1. Roll up a strip of lint tightly into a mass, 1 inch broad and $\frac{1}{2}$ an inch thick, trim the ends away with scissors till the mass is of a size to enter a posterior naris, then tie the wedge in the middle of a yard of whipcord previously doubled. If blood trickle down both nostrils, both must be plugged, and two such plugs must be prepared ; next, make two more similar rolls of lint, and tie these up with a short piece of silk or twine to prevent them from unrolling.

Step 2. Pass along the interior of the catheter a yard of twine, and draw its end through the eye of the catheter a few inches, then introduce the catheter through the naris directly backwards, not upwards nor downwards, because, when the patient is upright, the floor of the nose is nearly horizontal. When the catheter has reached the pharynx, the finger, or a forceps, must be passed through the mouth to catch the string hanging from the end of the catheter and bring it out of the mouth, where it is held while the instrument is withdrawn from the nose. The step is repeated in the other nostril if required.

Step 3. Next wash out the nostrils with a few syringefuls of ice-cold water, in which some tannin is suspended.

Step 4. Fasten the double string

Fig. 81.—Plugging the nares. Belloc's sound passed through the naris, and projecting at the mouth.

of the plug to the end of twine hanging out of the mouth (see fig. 81), and then draw the other end out through the nose ; this will carry the plug across the mouth to the pharynx, here the finger guides it over the soft palate and thrusts one of its ends into the naris, where the strings draw it tight. The plug for the anterior nostril is then put in place, and the strings tied tightly over it (see fig. 82). Thus the plug in front keeps the plug behind in place, and *vice versâ*. The end of string from the posterior naris, left hanging out of the mouth, must next be tied to the string of the anterior plug to keep it out of the patient's way, till wanted to withdraw the posterior plug, when that is to be removed. If blood run from both sides, the

Fig. 82.—Plugging the nares: the strings from the posterior plug tied over the anterior plug.

other anterior and posterior nares are stopped by a repetition of this operation.

This apparatus is very painful, and, if borne so long as a couple of days, should always be taken out then. If bleeding recur, which is very unlikely, fresh plugs must be introduced. Sometimes the posterior plugs are soaked in styptic solutions; this is a bad plan, because the bleeding part is not at the posterior naris, and the styptics increase the soreness the plugs themselves produce.

Another Plan.—Dr. Hamilton* has described a simpler method. A strip of linen 1 inch wide and 3 feet long is soaked in astringent fluid—glycerine of tannic acid, for example. The end is grasped by a pair of dressing forceps, and carried along the meatus to the posterior naris. About 1 foot of the strip is then rapidly passed into the nostril with the finger and

* " British Medical Journal," May 8, 1880.

thumb, making a solid mass, which is thrust as such to the posterior naris by the dressing forceps or a pencil. It should be distinctly felt occupying the posterior nostril by the finger passed through the mouth. A second portion should now be paid in quickly and pushed up the cavity of the nose as high as it will go. Lastly, the remainder should be quickly doubled into a solid mass and passed inside the anterior naris. Here the strip of linen remains until the end, loosened by mucous secretion, drops out. As the end hangs from the nostril it must be cut off bit by bit. But on no account should any traction be made on the strip : it should be allowed to come away of itself as the mucus softens and sets it free.

The introduction of the preliminary plug to the posterior naris may be facilitated by carrying a thread into the mouth with a catheter or Belloc's sound along the floor of the nose and attaching to it a loop of thread which runs freely round the end of the strip already folded into a solid mass large enough to fill the posterior naris, and secured in its shape. The loop of thread is then drawn backwards through the naris and mouth until the plug is fixed *in situ.* Then the loop is cut, and one end being pulled on, the whole thread slips away, leaving the plug fixed in the posterior naris.

Belloc's Sound.—Fig. 83 is a curved silver cannula like a female catheter, furnished with a long spring stylet, that arches round in a circle when thrust out of the cannula, and has a hole at the end to carry a thread. The long stylet can be unscrewed into two parts when not in use. The figure represents the instrument with the stylet ready for protrusion, and the same arching forwards after it is protruded.

The cannula is passed along the meatus till the end reaches the pharynx, then the stylet is protruded

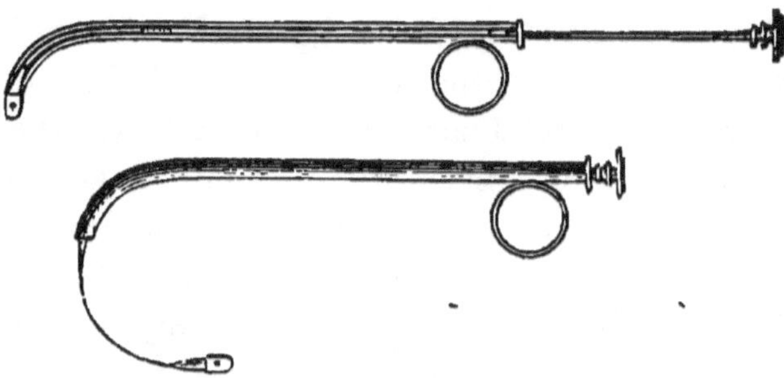

Fig. 83.—Belloc's Sound, for drawing a thread from the mouth along the meatus.

and arches forward till it reaches the teeth, when a thread is passed through the hole, and the stylet being withdrawn, the thread is carried with it into the pharynx and through the nostril, where it can be used to draw the plug into its place at the posterior naris.

Tooth Drawing.—A surgeon is frequently required to draw a tooth on emergency, and should be provided

Fig. 84.—Tooth Forceps.

with instruments (see fig. 84). Seven pairs of forceps and an elevator are sufficient for all he is likely to deal with. They are differently shaped for the different teeth, which vary much at the neck, the part grasped in the forceps.

For the operation the patient should be seated in a high-backed chair; the surgeon stands at his right side, holds the jaw with his left hand, while with the right he thrusts the beaks of the forceps

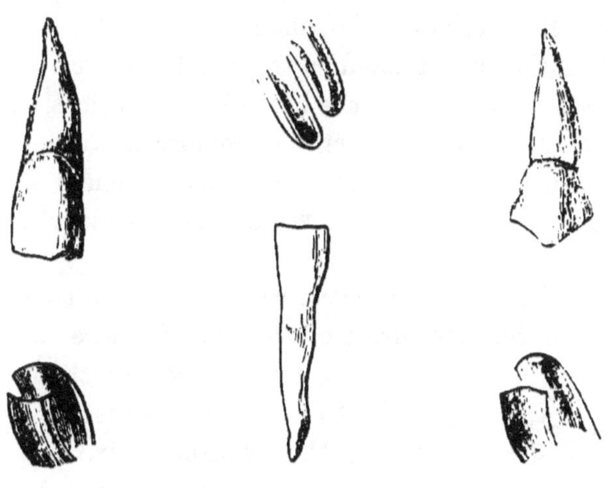

<table>
<tr><td>Fig. 85.—Upper in-
cisor tooth and
forceps.</td><td>Fig. 86.—Lower cen-
tral incisor and
forceps.</td><td>Fig. 87. — External
aspect of upper bi-
cuspid tooth, and
bicuspid forceps.</td></tr>
</table>

between the gum and the tooth on its lingual and buccal aspects; having reached the neck, he holds the tooth firmly, pushing it inwards and outwards with a rotary motion of the wrist (except for the molar teeth). Sudden tugs break the tooth and leave the fang behind; when loosened by rotation and lateral motion, or 'rocking,' the forceps readily lift the tooth out of its socket.

For the *upper incisors* the beaks of the forceps are straight, slightly hollowed inside, to give them hold of the teeth, and have crescentic edges (see fig. 85).

The *upper incisors* and *canines* can be drawn by the same pair, as the shape of these teeth at the neck varies to only a small extent.

For the *lower incisors* very narrow forceps are necessary. The beaks (fig. 86) should be curved at the joint sufficiently for the handles to clear the upper jaw (see fig. 88). The edge of the beaks is crescentic, similar to that of the upper incisors. These forceps are also very useful for removing roots, as their fineness enables them to sink between the stump and the alveolus with ease.

For the *bicuspids*, beaks with crescentic edges also are used, but the inside of the beak is more hollowed to fit the round neck of these teeth (see fig. 87). All the bicuspids can be drawn with the same pair, but it is convenient to have forceps bent at the joint to clear the upper jaw when extracting a lower bicuspid (see fig. 88).

Fig. 88.—Lower bicuspid forceps.

For the *upper molars* two forceps are required, one for each side of the jaw; the beaks of these are well hollowed to admit the crown of the tooth. The inner beak terminates in a crescentic border to fit the large internal fang (see figs. 89 and 90); the outer beak has two smaller grooves separated by a point that passes between the two external fangs.

In drawing these teeth the forceps should be thrust as high as possible and held firmly, while the fangs are loosened by moving the tooth from side to side, for from the multiplicity of fangs, rotatory motion is not available.

The *wisdom molars* are often difficult to seize from being almost buried in the jaw; as they resemble a bicuspid in shape, the bicuspid forceps (fig. 87) should be employed; if this fails to penetrate between the tooth and the alveolus, the narrow incisor forceps (fig. 86) can be driven up till it grasps the tooth. Not

Fig. 89.—Left upper molar tooth and forceps.

Fig. 90. — Right upper molar forceps.

unfrequently the fang of this tooth in the lower jaw is curved backwards and prevents extraction when the tooth has been loosened; this difficulty may be overcome by pushing the crown of the tooth a little backwards so as to tilt the fang forwards out of place.

When the molars are closely set, or the tooth to be extracted is overhung by its neighbour, it is often difficult to avoid tearing the gum extensively and even carrying away more than one tooth; tearing the gum is prevented by lancing it before applying the forceps, and slow and steady movements of the wrist usually prevent the latter accident, or the overhanging tooth may be filed away before the forceps are applied.

The *inferior molars* (fig. 91) have forceps, whose beaks are doubly grooved and pointed to enable them to seize the neck on each side between the two fangs.

In *raising stumps,* so much decayed that the forceps will not hold them, the elevator must be employed. This instrument (fig. 92), straight, pointed, and a little grooved at the point, is thrust down between the stump and the next tooth ; the jaw being then the fulcrum, the elevator is the lever to push *forwards* the fang ; when thus loosened it is easily lifted out. In working with an elevator there is some risk of thrusting the point through the alveolus, and wounding the tongue or floor of the mouth, hence it should always be guided and covered by the left forefinger. In removing the fangs of incisors the narrow forceps are most useful, and should it not be possible to penetrate between the fang and the alveolus, the alveolar border

Fig. 91.—Inferior molar tooth and forceps.

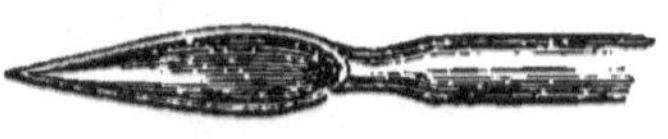

Fig. 92.—Elevator.

may be included in the grasp of the forceps and brought away with the tooth. The injury thus inflicted is very unimportant, and much pain is saved.

After the tooth is extracted the mouth should be well washed with warm water a few times, the attending bleeding being of no importance, except in individuals of hæmorrhagic diathesis, in whom measures should be at once taken to arrest the flow.

To stop a bleeding socket the alveolus must be well cleared of clots; then fragments of sponge, soaked in a solution of perchloride of iron, one part of the salt to three of water, should be packed into the cavity. A plug of cork is placed between the jaws, and a four-tailed bandage (see page 8) carried round the head to keep them firmly closed. Should this plan fail, the socket must be cleared again, and the wire of the galvanic cautery pushed well down to the bottom and then heated till it has cauterised the cavity.

The Laryngoscope is used to examine the interior of the pharynx and fauces, the larynx and the trachea. It consists (1) of a slightly concave reflector, with a central aperture, worn on the forehead, or before the eyes by means of a spectacle frame, of the observer. This throws a strong light to the back of the pharynx, whither is carried, by the hand of the observer, (2) a small circular mirror, set on a long pliant handle, and varying in size from that of a sixpence to that of a florin. This mirror reflects the light to that part of the interior opposite to which it may be turned, and the eye of the observer then perceives in the mirror an image of the part illuminated. If the face of the mirror look down, the larynx is depicted; if upwards, the naso-pharyngeal region, and so on. The use of the laryngoscope needs practice, but the difficulty in obtaining dexterity is not really great. For illumination sunlight is best, but failing that, a circular gas or oil flame will suffice; even a simple gas flame or candle will answer the purpose. The light must be placed behind the patient, level with his throat. Some clean napkins should be at hand for use.

To use the instrument: set the patient, with the knees together, upright, facing the observer, with the head thrown slightly back. Bidding the patient open his mouth and put out his tongue, the observer arranges the reflector to throw the light well into the pharynx, and gently seizes the tip of the tongue with his left thumb and forefinger in a napkin, steadying the left hand by resting the spare fingers against the patient's chin. In doing this no pull or strain on the tongue is made. Next the observer takes the laryngeal mirror in the right hand, between the thumb and fingers, like a pen, that it may be easily rotated or changed in position without the hand being itself shifted. He warms the mirror slightly by holding it over the lamp-flame, lest the breath condense on it and obscure the image, and tests it on the back of his left hand to be sure it is not *too hot*. This done, he slips the mirror to the pharynx, being careful to touch no part until the uvula is reached. This he gently pushes upwards, and thus brings his mirror exactly into the centre of the pharyngeal cavity. To see the larynx, the mirror is rotated till it reflects the light into that organ. At this moment the patient is bidden to take a deep breath, and a complete image of the larynx is obtained.

At the top of the image is seen the epiglottis; at the bottom, the two red prominences of the cartilages of Santorini poised on the summits of the arytenoid cartilages. Between these points are the aryteno-epiglottic folds of mucous membrane. At the lower part of the mirror, between the arytenoid cartilages, a fold of mucous membrane is seen during deep inspiration. This is the posterior wall of the larynx. The parts just enumerated form a frame, in which are seen the pearly white vocal chords, stretching like an inverted Λ, with the apex towards the epiglottis and the base towards the arytenoid cartilages. Still looking at the centre of the picture, below the vocal cords, the transverse rings of the trachea are seen, and, in favourable cases, even the bifurcation to the bronchi. Looking to the sides again, the red mucous surface of the false vocal cord is seen on each side, between the true vocal cord and the aryteno-epiglottic fold. During deep inspiration the vocal cords are widely separated, and the rima glottidis becomes lozenge-shaped. During phonation (as when the patient says "Oh") the cords lie close and parallel and the rima is a mere chink. They are then most strongly blanched.

It must not be forgotten that in the image the parts are reversed, so that the highest and apparently farthest part of the image represents the anterior portion of the larynx, and the lowest or nearest part of the image the posterior portion of the larynx. The left vocal cord also lies opposite the observer's right hand and the right cord opposite his left hand.

By skilful manipulation of the mirror a good view may be obtained of the base of the tongue, the pouch between the tongue and the epiglottis, the back of the tonsil and the pillars of the fauces; again, by slightly increasing the angle of the mirror and directing its surface upwards, an image of the back of the soft palate, the posterior nares and the naso-pharynx as far as the base of the skull (the basilar process of the occipital bone). To gain this the palate must hang loose and the base of the tongue be well depressed by a rectangular tongue depressor. The patient may also be made to emit a nasal sound. Nevertheless, in most cases, it requires great dexterity to obtain a good view of the back of the palate.

Nipple Shields and Artificial Nipples made of flexible ivory, vulcanised india-rubber, &c., are required when the nipple is chafed and excoriated by the child's sucking, especially if his mouth be attacked by thrush, as is usually the case. When the nipple is sore it should be well washed and dried after suckling, then covered with glycerine of starch or plastic collodion, and protected by a shield. If much inflamed it may be wrapped in lint dipped in alum water or solution of sulphate of zinc (one grain to the ounce), and deep chinks should be freely rubbed with lunar caustic. The breast should be regularly emptied by the breast-pump if the child's sucking gives much pain, lest the accumulation of milk in the ducts cause milk abscess.

Plugging the Vagina is employed in cases of rapid hæmorrhage from the womb, &c.

Apparatus.—1. A silk pocket-handkerchief. 2. A dry new fine sponge or pellets of cotton wool. 3. Silk thread. 4. A body roller or folded sheet.

The sponge should be cut into pieces the size of nuts; if the sponge is compressed it answers better. When prepared, the vagina should be cleared of coagula by a syringeful of ice-cold water; the handkerchief, unfolded and thrown over the right hand, is passed up the vagina till its centre reaches the os uteri, the borders and ends then project from the vagina. The interior of the handkerchief is next filled by firmly packing the sponge in bit by bit until the vagina is distended by the mass; the ends of the handkerchief are then tied together. The sponge swells as it absorbs the blood, and compresses the bleeding vessels by its distention. Before application it may be dipped in glycerine of tannin.

The abdomen and uterus are then supported by a body roller, or folded sheet, wrapped tightly round the hips and waist, while the patient, lightly clad, is kept quiet in a cool chamber.

When the plug has answered its purpose it is removed, by withdrawing the sponge bit by bit, and the vagina is washed with tepid water.

The kite's tail plug—Masses of cotton wool the size of a hen's egg are tied at two inches' distance from each other along a long string. When about a dozen are tied on, a speculum is introduced, and the first ball of wool is passed to the bleeding point and pushed firmly against it, and then another, and so on, until the vagina is firmly packed. An end of string is left hanging out of the vulva, whereby the plug may be removed when necessary. Each mass comes away successively with ease as the string is pulled out of the vagina.

Inflatable india-rubber tampons are now employed for this purpose, but they are not always at hand, and are likely to get out of order if seldom used.

Injecting the Urethra often fails from the inefficient mode in which it is done. The syringe employed should be short enough to be worked easily with one hand, and need not contain more than two or four tea-spoonfuls, as that amount distends the urethra. One of such a size is easily worked by one hand. The opening through the nozzle should also be wide, that a forcible stream may be injected into the urethra.

The patient should fill the syringe, then place on a chair or stool before him a chamber-pot, and make water to clear out the discharge collected in the urethra. He then inserts the slightly bulbous nozzle into the meatus urinarius, and grasps the sides of the glans with the left forefinger and thumb to close the mouth of the passage. The right thumb next presses down the piston slowly, so that the whole of the injection passes into the canal and distends it. Keeping the meatus shut with his left finger and thumb, the patient lays down the syringe and rubs the under part of the penis backwards and forwards, that the injection may be forced into all folds or follicles of the mucous membrane. Having thus occupied about thirty seconds, he releases the mouth of the passage, when the fluid is ejected sharply into the vessel placed ready to receive it. This rapid ejection is a test of the proper performance of the operation.

In counselling the use of astringent solutions, the surgeon should always caution the patient not to employ one that produces smarting that is severe, or which lasts more than a few minutes after injection. If it causes much pain, the solution is too strong.

Catheters and Bougies. — Silver catheters are made in sizes, increasing from No. ¼ to No. 12, the first having a diameter of 0·06 inch, the latter 0·25 inch. Larger ones than these are seldom employed. Each catheter is fitted with a wire stylet.

Fig. 93.—Silver catheter.

The curve preferred by different surgeons varies much; the one most generally adopted is that used by Sir Henry Thompson. This curve forms one-fourth

Fig. 94.—English flexible catheter.

of a circle of 2½ inches' diameter, and is that most closely agreeing with the curve of the deeper part of the urethra.

The flexible catheters of woven silk are of many

Fig. 95.—French bulbous-ended catheter.

kinds; the English so-called 'gum elastic' (fig. 94), the French black flexible (fig. 95), and the vulcanised india-rubber (fig. 96) catheters being the three varieties

Fig. 96.—Vulcanised india-rubber catheter.

most generally employed. English flexible catheters should be kept on stylets well curved at the last

3 inches, that, when the stylet is withdrawn, for the catheter to be passed, the latter may retain sufficient curve to glide over the neck of the bladder easily.

Vulcanised india-rubber catheters are now made with projections at the eyes to prevent them from

Fig. 97.—The self-retaining catheter.

slipping out of the bladder. Fig. 97 represents one pattern of these *self-retaining* catheters.

Sounds are solid, being of steel, plated or gilt, or simply polished. Their curve varies, and is generally 20 or 30 degrees more obtuse than that of the catheters. Their flat handles are set across the butt so that the upper surface faces the same way as the point of the instrument, and consequently always indicates the direction to which the point is pressing during its passage along the canal.

Bougies are made of the same materials as the flexible catheters, also of whalebone, catgut or silkworm gut ; they are kept straight, and the more supple they are the better. The black bulbous-ended bougies (*bougies olivaires*) are the most useful variety for dilating the urethra.

Bullet-Headed Bougies (*Bougies à boule*) are used for exploring the urethra in cases of gleet, where the discharge is often kept up by a stricture or a tender patch of chronic inflammation of the mucous membrane. They are made of metal, or of black gum mounted on a very flexible leaden wire ; both kinds are useful. The stem of the instrument is slender, no bigger than a No. 2 or No. 3 bougie of the English scale; the end terminates in an egg-shaped bulb of any required size. These bougies are most useful from No. 4 to No. 20 of the English scale, corresponding to Nos. 10 and 30 of the milli-metrical scale. The stem should be marked with

Fig. 98.—The bullet-headed bougie.

white rings an inch apart, the fifth inch from the bulb having a ring broader than the rest, so that when the instrument is passing over a tender part, or is arrested by a stricture, the distance of the impediment down the urethra can be at once estimated. In withdraw-ing the instrument, the wide base of the olive shows the exact position and length of those strictures which are not too narrow for the olive-head to slip by, for it is nipped by the stricture and released as soon as the narrowing is passed. By using instruments large enough to fill the normal urethra, an induration beneath the mucous membrane can be detected in its earliest stage before it has produced symptoms diagnostic of stricture.

Rigid instruments have one advantage over flexible ones, in that their points can be guided by the sur-geon; the points of flexible instruments cannot be directed, hence the introduction of the latter into a stricture is less easily managed, consequently bougies with various kinds of points should be kept. But flexible instruments cause far less irritation than rigid ones, and should always be employed instead of the latter when possible. With patience and practice, much of the difficulty attending their introduction is overcome. The French bougies, with tapering ends and bulbous points, slip more easily through a stricture than instruments having the same diameter throughout, and bougies with fine tapering points can sometimes be introduced where others fail.

Passing Catheters.—In passing instruments along the urethra the conformation of its interior should be borne in mind. From the meatus to the triangular ligament, the normal urethra, when gently stretched, becomes a straight tube ; having, nevertheless, just within the meatus, a pouch in the roof, the lacuna magna, where the point of the instrument may catch if not turned downwards. At the bulbous part the urethra enlarges in capacity by having a slight downward curve in its floor, just before the triangular ligament is reached. In this depression, the beak of the catheter is apt to sink below the level of the passage through the ligament, which is always a fixed point. Beyond the triangular ligament the urethra curves gently upwards, has a floor beset with irregularities, in which the point of the instrument easily catches, if not raised as it passes along the curve.

A solution of carbolic acid in olive or almond oil, 1 part to 16, should be the ordinary lubricating oil ; and no instruments should ever be laid aside before they have been thoroughly cleaned in hot water and wiped dry.

A Silver Catheter is passed most easily while the patient is in a horizontal position, with the shoulders low and the thighs separated. The surgeon stands on the left side of the patient, and holds the catheter, previously warmed and lubricated with carbolic oil, lightly between the thumb and two first fingers of the right hand, the beak downwards and the stem across the patient's left groin. Then taking the penis between the middle and ring fingers of the left hand, the palm being upwards, he pushes back the foreskin with the thumb and forefinger, and steadies the meatus while introducing the beak of the catheter. This done, he draws the penis gently along the catheter as the point is lowered to the perinæum, but without raising his right wrist until the instrument has travelled 5 or 6 inches along the passage and reached the triangular ligament. The surgeon then carries his right wrist to the middle line of the patient's body, and while pushing the point onwards, raises the hand round a curve till it again sinks between the patient's thighs. When the bladder is reached he withdraws the stylet that the urine may escape. Three points of difficulty are usual in passing catheters—the lacuna magna just within the meatus, the triangular ligament, and the prostatic part of the urethra just before the bladder is reached. The first is escaped by keeping the beak along the floor of the urethra for the first two inches; the second is best avoided by raising the wrist as the instrument passes the triangular ligament, and directing the beak against the upper surface of the urethra, lest, being in the enlarged bulbous part, it sink below the opening in the ligament; the third difficulty is overcome by depressing the hand well as the point approaches the bladder.

To pass the catheter in the upright position, the patient is placed against a wall or firm object, with his heels eight or ten inches apart and five from the wall, that he may rest easily during the operation. The surgeon seats himself opposite the patient and grasps the penis with the two middle fingers of the left hand, the palm upwards; he next exposes the meatus with the thumb and forefinger; and, his right hand holding the catheter by its middle obliquely across the left side of the patient, he draws the penis on to the instrument till the triangular ligament is reached. He then carries the shaft of the catheter to the middle line, and, holding it by its end, brings the right hand downwards and forwards, to carry the point upwards over the obstruction at the neck of the bladder.

The operation should be done slowly and with great gentleness, giving the urethra time " to swallow the instrument," as the French surgeons express it. Hasty or forcible movements tend to thrust the point against the wall of the urethra, where it hitches, if it does not penetrate and make a false passage. However easy the introduction may have been, the withdrawal of the catheter should be always done slowly to avoid giving pain to the patient.

When the canal suddenly contracts, as from a stricture, the point of the sound often stops at the obstruction; by withdrawing the instrument a little, and diverting its point to another side or along the upper part of the urethra, a point where the obstruction is less abrupt will often be found to let the catheter glide into the stricture. The floor of the urethra should always be avoided, as false passages nearly always branch off from the floor close to the stricture.

Difficult narrow strictures are most easily overcome

by injecting a drachm of warm carbolic oil into the urethra, and then passing fine black silk or whalebone bougies (*bougies filiformes*) along the urethra. These, from their fineness (their diameter is only $\frac{1}{4}$ or $\frac{2}{3}$ of a millimetre, about $\frac{1}{100}$ inch), are very apt to catch in false passages : if so, the bougie should be left engaged in the false passage, and held in the left hand while another bougie is passed along the urethra : if, in its turn, this one gets into a false passage, it also should be left, and a third passed ; and so on till all the false routes are occupied, or a bougie enters the stricture and reaches the bladder, which is known by the readiness with which it will pass backwards and forwards. The other bougies should then be withdrawn, and the bougie which has passed the stricture be tied in for twenty-four hours, until the passage is sufficiently dilated to allow a small catheter to replace it. If the patient is not suffering from retention of urine, there need be no anxiety about evacuating his bladder, as urine will find its way alongside of the bougie when he attempts to make water. In passing to relieve retention, No. $\frac{1}{2}$ English flexible catheter should be used instead of bougies : but when the stricture is too narrow for these, a bougie may still be tried, as the urine will generally dribble by the side of the bougie with sufficient rapidity to relieve the patient.

English flexible Catheters should be kept on stylets curved as represented in fig. 94, that the first 3 inches of the instrument, when the stylet is withdrawn, may retain sufficient curve to ride over the impediment at the neck of the bladder. In warm weather, after being oiled, they should be dipped in cold water just before using, to render them a little stiffer, and less likely to lose their curve while traversing the urethra. Any curve may be imparted to the English flexible catheter by first softening it in hot water, and then holding it in the required curve while it is dipped in cold water to set it.

They may also be passed while the patient lies or stands, and the movements are the same as for the silver catheter.

Bulbous-ended or probe-ended Catheters and Bougies (*Bougies olivaires*) are always straight; their suppleness, their tapering ends, and their smooth rounded point enable them to glide along the urethra, and to accommodate themselves readily to the windings of the passage; for which reason they are the easiest to pass both for the patient and the surgeon. In passing them they are slightly warmed if the weather is cold, to restore their flexibility, and gently pushed along the canal till the bladder is reached.

Fig. 99.—The Coudé catheter.

Elbowed or Coudé Catheters.—These have a fixed abrupt curve of $1\frac{1}{2}$ right angles, as represented in the figure, $\frac{1}{2}$ or $\frac{3}{4}$ inch from the point. They are most useful in cases of enlarged or inflamed prostate. Bi-coudé catheters have two such curves.

Vulcanised India-Rubber Catheters (figs. 96 and 97) are used when the bladder is to be kept empty; their suppleness renders them unirritating, and as phosphates crust on them very slowly, they may be worn for a week without being changed.

They are easily passed by threading them on a long stylet with the appropriate curve, and tying them firmly to it. The stylet is withdrawn after they are passed.

Recently a new material, "celluloid," has been patented as a material for catheters. It is very smooth, indestructible, and supple when hot, retaining, as it cools, any curve given while hot.

To pass a Catheter on the Female.—The patient may lie on her side or on her back; if on her side, the knees should be well drawn up; if on her back, the thighs must be somewhat separated. Before introducing the catheter, a wine bottle or narrow-necked bed urinal should be placed in the bed ready to receive the urine. If the ordinary slightly curved female catheter be not at hand, a No. 7 or 8 flexible one does just as well.

Having oiled the instrument, go to the patient's back, and take the catheter in the right hand if she lies on her right side, and in the left hand if she lies on her left side ; if she lies on her back, go to either side and take the catheter in the hand nearest her feet. Hold the stem of the catheter in the palm, so that the beak lies against the tip of the forefinger, while the thumb and second and third fingers grasp the stem. Then passing the hand under the bed-clothes, seek the buttock; from that pass the forefinger to the perinæum, and let it enter the vulva, keeping the back of the finger against the posterior part, then pass it between the nymphæ to the entry of the vagina.

This is known by the tip of the forefinger being lightly grasped, unless the vagina is very wide. Keeping the finger just within the vagina, feel for the arch of the pubes in front ; having found this, withdraw the tip of the finger slightly from the vagina : in doing this, it will strike a small projection of mucous membrane hanging just at the anterior margin of the entry. Keep the finger steady against this projection, while the other hand pushes the catheter gently onwards, which then rarely fails to enter the urethral opening which lies close above the projection of mucous membrane. Having penetrated the urethra, arrange the catheter in the receptacle for the urine, and push the instrument into the bladder.

To Wash out the Bladder.—*Apparatus.*—1. A flexible catheter ; Nos. 12 or 13, and of single " web," so as to have a large interior channel through which mucus can escape, but a smaller size must often be employed. 2. A caoutchouc bottle, holding six ounces, and fitted with a tapering nozzle and stop-cock. (Fig. 100).

Fig. 100.—Elastic india-rubber bottle for injecting.

During the operation the patient should stand, if possible, as the mucus is thus more easily cleared from the bladder. The surgeon first fills his bottle completely with tepid water, that no air may remain ; then directing his patient to stand against a wall or some firm object, passes the catheter and draws off the urine. He next inserts the nozzle into the catheter, and, turning the cock, compresses the bottle slowly until two or three ounces of water have run into the bladder; this he lets escape by removing the bottle for a minute, and then repeats his operation till the water returns clear, but is careful not to exhaust the patient's strength. Three or four small injections wash the sediment and mucus from the bladder as quickly, and with far less fatigue or risk of spasm than a prolonged flow of water through a stiff double current catheter. In this way the bladder may be washed twice or thrice daily to the great comfort of the patient.

Solutions of nitrate of silver, carbolic acid, alum, &c., in the proportion of 1 part to 100, or to 50 of water, can be used instead of water for this purpose.

To Tie in a Silver Catheter.

Apparatus.—1. A few yards of tape ¼ inch wide. 2. A roller. 8. A spigot of wood; or, 4, A yard and a half of fine india-rubber tubing.

A 2-inch roller is tied round the hips; from this, on each side, a

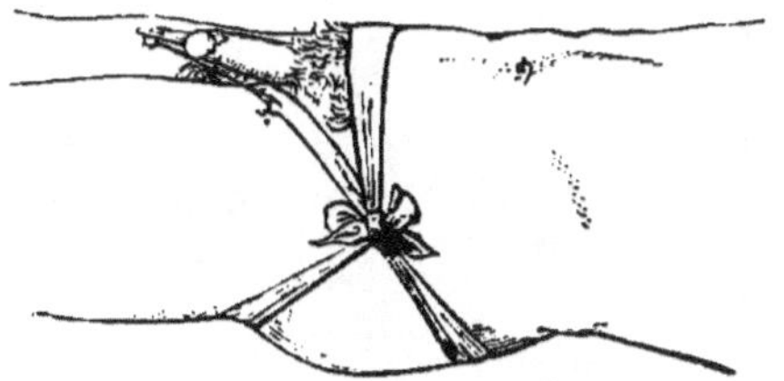

Fig. 101.—A silver catheter tied in the urethra.

tape is passed round the thigh at the groin, and fastened before and behind to the roller round the hips (see fig. 101) ; a narrow tape run through the rings of the catheter connects them with the loops in the groins. The tapes are tied short enough to prevent the catheter slipping out; a yard or two of fine india-rubber tubing, fixed on to the end of the catheter, conveys the urine to a pan under the bed, and keeps the bed dry ; or a spigot of wood fitted to the catheter may be inserted, for the patient to draw out when he desires to void his urine.

To Tie in a Flexible Catheter.   (Fig. 102).

Fig. 102.—A flexible catheter tied in the urethra ; the string fastened behind the corona glandis, and concealed by the foreskin.

Apparatus. — 1. A piece of soft twine, or knitting cotton, about 15 inches long.

A catheter is first passed into the bladder, and the urine runs off. The catheter is then gently withdrawn, till the stream ceases, that the end of the instrument may remain just without the neck of the ·bladder. The string should be tied round the catheter $\frac{1}{2}$ an inch from the meatus, its ends gathered together and tied in a knot about 1 inch farther on. The foreskin is then drawn back, the ends passed beneath the glans, and tied round the penis behind the corona ; the superfluous string is snipped off, and the foreskin brought forward. The catheter is cut off obliquely $\frac{1}{2}$ an inch beyond the string, and then stopped with a spigot, direction being given to the patient to withdraw the spigot, and push the catheter a little further in when he wants to make water.

To Tie a Patient in Position for Lithotomy.

Apparatus.—Two bandages, each 3 yards long and 2 inches wide, of calico or of saddle-girth, with tapes sewed on the ends.

The patient is laid on his back, a slip-knot made in the middle of the bandage and passed over the wrist ; the hand is then made to grasp the foot, the thumb above, the fingers under the sole (fig. 103) ; one end of the bandage is carried behind and inside the ankle to the dorsum of the foot, where it meets the other end passing in front of the ankle. The ends are then carried under the sole, brought up again, and carried round the hand and foot till they can be secured by tying the tapes in a double bow over the back of the hand. It is essential in laying

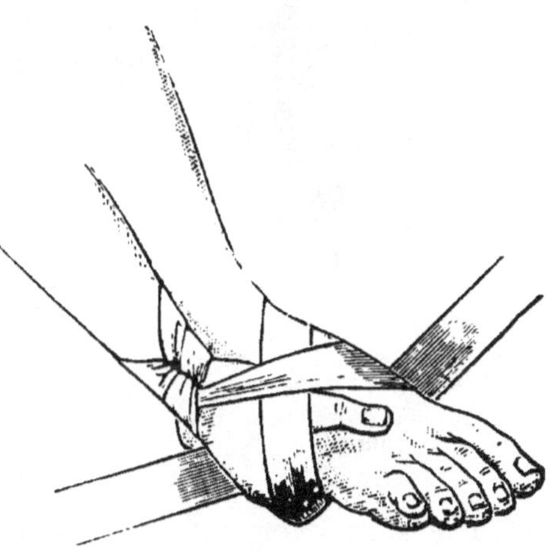

Fig. 103.—Tying for lithotomy.

on this bandage that each turn be tight and no slack left. The assistants who hold the patient stand facing the surgeon on either side of the table, in order to keep the thighs widely and evenly separated during the operation. The one in charge of the right limb passes his left arm round the thigh, and grasps the leg below the knee, while his right hand holds the everted foot, and thus steadies the abducted limb— he who holds the left limb, mutatis mutandis, does the same.

Brown's Dilatable Tampon is an ordinary lithotomy tube surrounded by an india-rubber bag (see fig. 104), which can be inflated through a small india-rubber tube. In the sketch the bag is represented in its collapsed form, B, and the dotted oval, C, marks the extent to which it can be puffed out. By the four-tailed string the tube is fastened to the hip bandage applied as directed for securing a silver catheter. This method is excellent for preventing hæmorrhage after lithotomy.

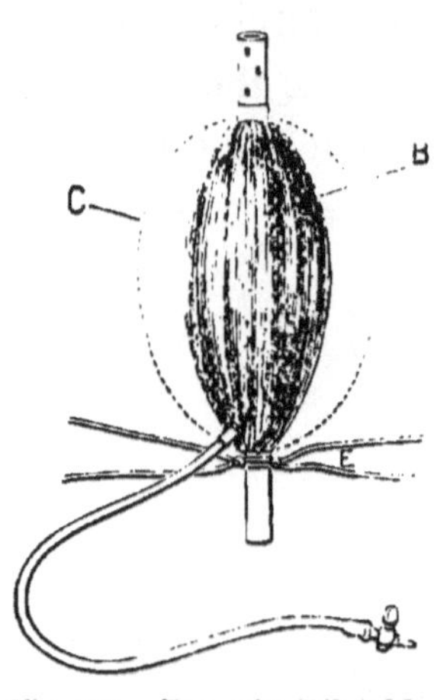

Fig. 104.—Brown's Dilatable Tampon.

Bedsores are best treated by great cleanliness, and by washing the skin exposed to the discharges with spirit of wine every day.   The pressure of the skin over the sacrum or trochanters is prevented by a ring of soft thick felt, covered on one side with adhesive plaster, and applied like a corn plaster *round* the prominent bone.

In addition to these local applications, the pressure of the body should be evenly distributed over its under surface by placing the patient on a water-cushion, or, better, on Arnott's water-bed, or, where possible, by frequent change of posture.

Arnott's Floating Bed.—In the hydrostatic or floating bed of Dr. Arnott, the patient floats on the surface of a trough of water, into which he sinks until he has displaced his own weight of the fluid ; the floating apparatus, or raft, so to speak, on which he lies, being a sheet of water-proofing, and a thin mattress or folded blanket. The bed consists of a trough running on large castors, about 8 feet long, 2 feet 8 inches wide, and 1 deep, with a tap at the bottom for letting out the water, and a spout in one corner to fill it by. Over the top a macintosh cloth is spread, its edges being firmly nailed to the margin of the trough, but the cloth is left slack enough to float easily on the surface of the water when the trough is partly filled. This slackness is requisite to allow the water displaced

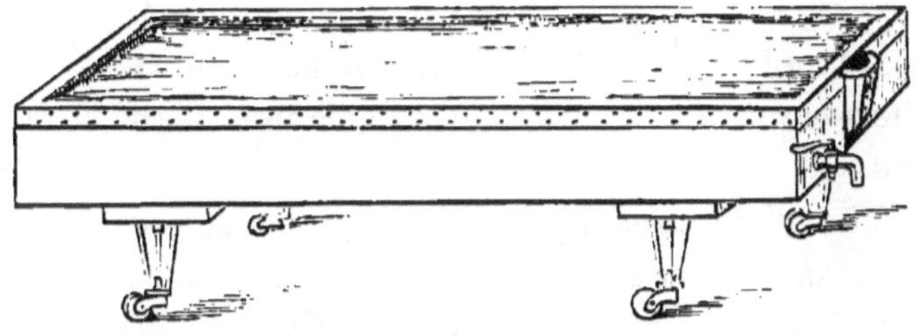

Fig. 105.—Water-bed.

by the weight of the patient's body to rise up round him without tightening the cloth, or the floating principle of the bed is not carried out, and the pressure of the patient's weight not evenly distributed over his body (see fig. 105). Three or four blankets are laid evenly over the macintosh, and these again are protected from the moisture of the patient by a macintosh under-sheet. If a mattress is used, it must be very thin and supple, to let the surface of the water adjust itself to the patient's body and receive the pressure evenly. The temperature of the water employed to fill the bath should be about 60° Fahr.

Water-Cushions are made of stout macintosh cloth, half or two-thirds full of water, and laid on the

mattress beneath the blanket and sheet (see fig. 106).
They are more portable than the water-bed, but they

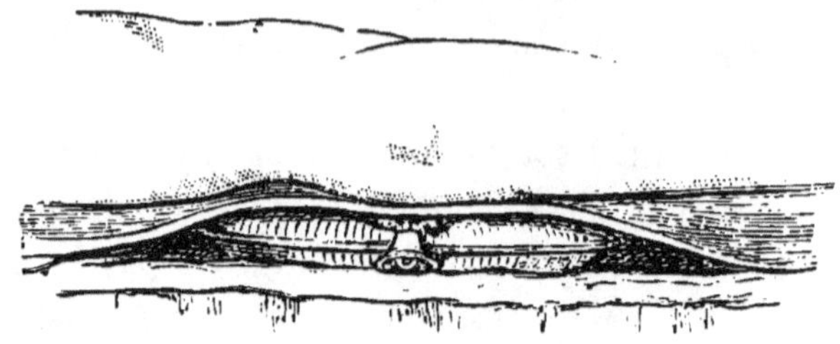

Fig. 106.—Water-cushion.

are simply soft pillows, and do not counter-balance
the weight of the patient in the manner of the floating
bed.

The *Water-Mattress*, constructed on the same principle as a water-cushion, is employed as a convenient
substitute for the water-bed, though less efficient.   It
is laid on an ordinary mattress, and covered by two or
three folded blankets, and a macintosh under-sheet.

Air-Cushions are also most convenient stuffing pads
for invalids. They are made of impermeable macintosh
cloth, and are far lighter than water-cushions.

The Coin-Catcher is an ingenious contrivance
for removing a coin or other foreign body impacted in

Fig. 107.—The coin-catcher.

the gullet.   It consists of a flexible whalebone rod
(see fig. 107) tapering slightly towards one end, to

which is attached a double broad shouldered metal
cradle, working freely on a pivot so as to project from
either side of the rod. This is passed below the foreign
body, which on the instrument being withdrawn, is
caught by the projecting wing and easily brought up.
A sponge is attached to the other end of the rod to
sweep out fish-bones, &c., when in the gullet.

The Stomach-Pump is used for emptying the
stomach, or for injecting fluid food when patients
refuse to swallow.

It consists of a brass syringe holding 4 ounces, of
which the nozzle is connected with two tubes, one at

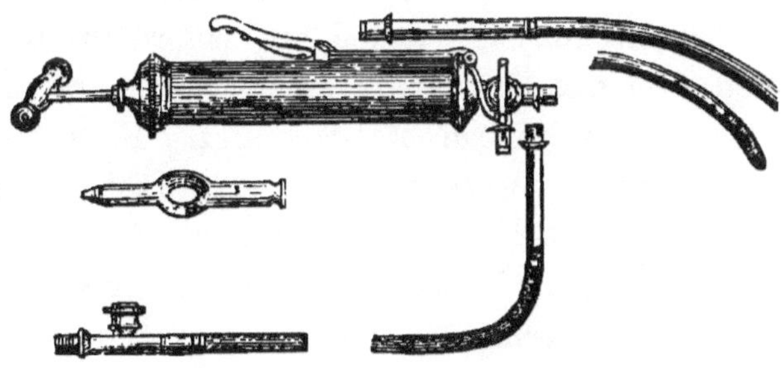

Fig. 108.—The Stomach-pump.

the end, the other at the side. The passage through
them is directed by a valve which is governed by a
lever lying on the barrel (see fig. 108). When the lever
is at rest, the current passes in and out of the syringe
by the lateral tube ; when depressed, by the direct
tube. The elastic tubes with smooth nozzles, about
2 feet long, are fitted to the syringe. There is also a
gag of hard wood, having a hole in the middle, through
which the tube passes on its way to the stomach, to
protect it from the patient's teeth.

When the pump is employed to remove the contents of the stomach, two washhand basins are placed at hand, one empty, one full of tepid water. The patient is seated in a high-backed chair to steady his head. If he resist the operation, one assistant holds his hands, while a second screws the small end of the gag between the teeth and forces open the mouth, across which it is then easily fixed. The flexible tube, being well oiled, is next passed across the pharynx and down the gullet slowly and cautiously, without staying for any effort of vomiting it may induce; when about 20 inches are passed through the gag the nozzle has reached the stomach. It is always prudent to wait until the patient has expired air, before any fluid is injected to make sure that the tube has not passed into the larynx instead of the gullet. First, two or three syringefuls of water are injected into the stomach; then, removing the second tube from the basin of water to the empty basin, the action of the syringe is reversed, by pressing on the lever as the piston is raised, and letting it fly up when the piston is depressed. Thus two syringefuls may be withdrawn, then fresh water is again injected and withdrawn until the contents of the stomach are removed and the water returns clear. Precaution must be always taken not to exhaust from the stomach before water is injected, lest the coats of that organ be injured by being sucked against the nozzle of the tube.

If desirable, antidotes may be dissolved or suspended in the water injected. When the pump is used for feeding patients, one or two pints of beef tea, eggs beaten with milk or wine, Liebig's soup, &c., are the kinds of food suited for the purpose. Each time the pump is used it should be thoroughly cleaned by syringing through it plenty of warm water, and the tubes must be unscrewed to wipe the joints carefully.

To wash out the stomach by syphon action.—A method of treatment employed in some forms of dyspepsia. *Apparatus.*—(1.) To a tube, of the material of which œsophageal bougies are made, slightly bulbous at the end, about ½ inch in diameter and 20 inches long, is fitted a length of india-rubber tubing 18 inches long ; at the end of this a glass funnel or receiver capable of holding a pint, is adjusted. To the nozzle of the funnel is a stop-cock. (2.) Two quarts of tepid water containing ℥ij of carbonate of soda. When about to be used, the stiffer end is moistened or lubricated with vaseline, and the patient, sitting upright, passes the free end to the pharynx, and, while keeping up a series of swallowing movements, rapidly thrusts the tube down his gullet. When all the stiffer portion has passed into the mouth, the patient stops. The end of the tube lies then along the greater curvature of the stomach. He next charges the funnel with alkaline water, and then raises the funnel above his head until the fluid has almost all subsided into the stomach. At this moment he lowers the receiver below the level of the stomach, and over a basin. The fluid immediately returns into the receiver, and is emptied into the basin below. A larger bulk of fluid runs out than had entered, being mingled with the residue of digestion, &c.

The operation is repeated as often as is necessary, until the alkaline water is returned in a limpid state.

The first injections are tepid, the latter ones cold, because they cause slight contraction of the stomach. The first injection may be five or six ounces of milk. This operation is usually performed once daily, but in extreme cases may be employed twice in twenty-four hours.

Besides merely washing the stomach, this plan of introducing the stomach tube may be employed for feeding patients whose appetite is insufficient. In such cases the food is carefully prepared by cooking, and is usually of the consistence of gruel or purée.

Fine œsophageal tubes, varying in size from No. 12 to No. 6 of the English catheter scale, are manufactured to use in feeding patients with stricture of the gullet.

Transfusion of Blood.—The points of greatest importance in performing this operation are :—

1. That the supply of blood come from a vigorous adult.

2. That the transfer be made within two minutes of the blood's escape from the vein of the blood-giver.

3. That, to prevent coagulation, the blood should pass over as small a surface, and suffer as little exposure as possible in transit.

4. Care must be taken to prevent air from entering the vein with the blood.

The apparatus described below is that devised by Dr. Graily Hewitt, and depicted in the Obstetrical Society's Transactions for 1864, page 137. It consists of a glass syringe holding two ounces (fig. 109), with a piston easily attached and removed ; its nozzle is curved and fits the mouth of a cannula of silver. The nozzle of the syringe is provided with a little stopper attached by a chain ; a stylet likewise fills the cannula, to be withdrawn when the blood is injected through the latter.

The success of the operation depends in great measure on the rapidity with which it is performed, and requires the aid of two assistants that the various steps may follow each other as quickly as possible.

Apparatus. — 1. Syringe, cannula and stylet. 2. Lancet. 3. Scalpel. 4. Forceps. 5. Three yards of tape, one inch wide, and lint. 6. A silver wire suture. 7. A basin of cold water. 8. Brandy and Sal Volatile.

Step 1. See that the piston-rod works properly in the syringe, and that the instrument is fit for use ; then place

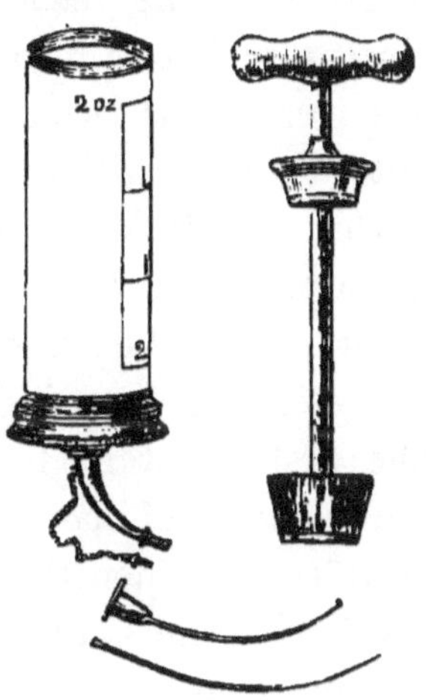

Fig. 109.—Graily Hewitt's syringe for transfusion of blood.

it in the basin of cold water, with the cannula to lie till wanted.

Step 2. Place the blood-giver on a couch or easy chair in the same chamber, but so that he cannot see the patient, lest he faint and his blood consequently flow feebly. Tie up the arm as for venesection ; lay ready the lancet, and direct the assistant, in charge of the blood-giver, to keep his thumb on the vein when it is opened, that the flow may be checked when blood is not required.

Step 3. Place a tape round the arm of the patient above the point for injection, and another below it at a convenient distance, and lay bare a vein (usually the median basilic) for an inch and a half of its course; holding the vein by the forceps, make a slit with the scalpel and introduce the cannula, which is then intrusted to the second assistant. The stylet is withdrawn, and a minute drop of blood escapes through the cannula, showing that the point has been properly introduced into the vein. The assistant replaces the stylet and slackens the upper ligature, while the surgeon proceeds to fill his syringe.

Step 4. The surgeon, going to the blood-giver, makes a large opening in the vein with a lancet, or if the first assistant be a surgeon also, he may do this while the chief operator is preparing the vein of the recipient. When the vein is open and the blood flowing freely, the barrel of the syringe is inverted over it and filled with blood; when full, the nozzle is stopped by the plug and the piston attached while the syringe is carried to the recipient.

Step 5. This being reached, the plug is pulled out, the nozzle inserted into the cannula, and the blood *slowly* injected by depressing the piston gently, but without quite emptying the syringe. A minute should be spent in injecting one ounce and a half, and a pause of five minutes ensue before a second supply is introduced. This interval may be employed in cleaning the syringe, &c., and procuring a fresh supply of blood; 3—4 ounces of blood are usually sufficient, but 10 ounces have been injected on some occasions. The perturbation of the supplier (generally a near friend of the recipient), often renders it necessary he should drink freely of brandy and water, that the blood flow forcibly when required.

Flexible tubes with nozzles to fit the veins of the donor and recipient, with an elastic injecting ball, are also contrived for the transfusion of blood, but the india-rubber, if not constantly in use, becomes brittle, and consequently cannot be trusted for an emergency that seldom occurs. There are also difficulties in bringing the donor and recipient as near to each other as the immediate flow of the blood through the india-rubber tube requires.

Step 6. When sufficient blood has been introduced, both patients' wounds are dressed, as after venesection (see page 20), the long incision of the recipient being closed by a point of suture under the pad.

Transfusion by Roussel's Apparatus.

Part of this apparatus consists of an arrangement for filling the tubes which afterwards convey the blood

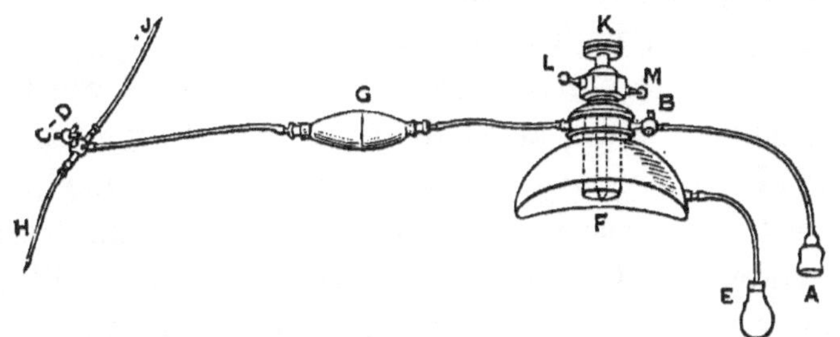

Fig. 110.—Roussel's Transfusor.

to the recipient, with tepid solution of bicarbonate of soda, in order to warm them and to clear them of air. Another part consists of a receiver, F, fitted by atmospheric exhaustion to the arm of the blood-giver over the vein which is to be punctured at the right moment by a lancet placed in the receiver. The third part is the Higginson's pump, G, which drives the blood along the tubes into the vein of the recipient.

Apparatus—1. Roussel's india-rubber and vulcanite apparatus (fig. 110). 2. Scalpel. 3. Dissecting forceps. 4. Two bleeding tapes. 5. Sponges. 6. Lint and diachylon plaster and collodion. 7. A suture. 8. A quart of tepid water, in which half a teaspoonful of bicarbonate of soda is dissolved. 9. A second basin of warm water in which to plunge and wash the apparatus before it is used. 10. A glass of warm brandy-and-water, sal volatile, and other restoratives.

Step 1. Prepare the apparatus by pumping warm water through it while it lies immersed. Get ready the several articles required.

Step 2. Bring the blood-giver close to the bedside of the recipient, and seat him comfortably.

Step 3. First tie a bleeding tape around the blood-giver's arm to obstruct the flow through his veins at the elbow. Next, apply the receiver, F, to the arm of the blood-giver, so that its centre, through which the lancet will protrude, overlies the median basilic vein. Compress the ball, E, so as to empty it of air. While holding E squeezed together, press the receiver, F, firmly down on the blood-giver's arm. Relax the pressure on E, whereby the air in F is withdrawn, and a vacuum produced therein.

Step 4. Turn the two-way stopcock, C D, so that the third small knob on it points towards the nozzle,

H, which is to be inserted into the recipient's vein ; plunge the vulcanite bell, A, into the soda-solution, and work the Higginson's bulb, G, until the whole apparatus, receiver, F, tubes, and pump, G, are filled with soda-solution, and it flows out through the open nozzle, H. Then turn C D to stop the flow.

Step 5. Expose an inch of the median basilic vein in an arm of the recipient, and, puncturing it, insert the nozzle, H, into the recipient vein. Intrust the nozzle, H, and the recipient's arm to an assistant. (This assistant, if he is competent, may perform the exposure of the vein while others are fitting the apparatus to the blood-giver, and thus save time.)

Step 6. Set the little German-silver knobs, L and M on K, diagonally across the direction of the blood-giver's vein, so that the lancet projecting below K, may penetrate the vein obliquely. Next, press down smartly the knob, K, and thereby puncture the vein beneath it. Blood escapes from the vein, and gradually replaces the soda-solution in the receiver, F. Shut off the stopcock, B, through which soda-solution reaches the apparatus. Next, turn the stopcock, C D, and work the pump, G, so that soda-solution flows out of the nozzle, J. When blood instead of soda-solution issues from J, turn the stopcock, C D, quickly, so that the current is diverted from J to H. Continue to pump *slowly* through H, which is steadily held in the recipient's vein until the requisite amount of blood is transfused from the giver to the recipient.

Step 7. Remove the receiver from the bloodgiver ; tie up his arm as after venesection (see p. 21). Tie up the arm of the recipient likewise, after bringing the edges of the skin-wound together with a stitch and collodion.

When the apparatus has been used, put it at once into warm water and pump through it plenty of water and of soda-solution to cleanse it thoroughly. Then unscrew the several parts and wash them thoroughly. Then dry the whole carefully, pulling out K in order to clean and dry the lancet properly. Let the apparatus lie before a fire to get perfectly dried before it is put away in its case. Neglect of this instruction will render the instrument unfit for use at the next emergency.

If the bloodgiver be nervous or faint, he must be freely plied with restoratives.

Tourniquets.—Tourniquets are of several kinds. When hæmorrhage has to be temporarily arrested,

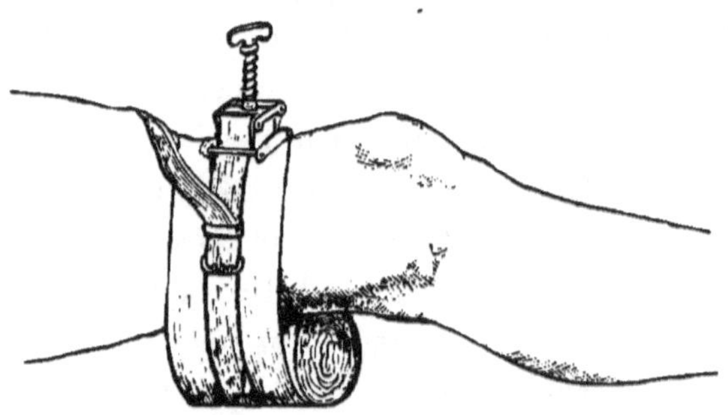

Fig. 111.—Petit's tourniquet applied to the popliteal artery.

that of *Petit* (fig. 111) may be used. It consists of a strap of stout webbing and buckle, that can be rapidly tightened by a few turns of a screw. To use this tourniquet : lay a roller over the artery as a pad and make it steady by carrying the end once or twice round the limb ; then pass the strap over the roller, keeping the buckle about 2 inches away from the screw and the screw on the anterior or outer aspect of the limb, away from the pad, lest that be displaced when the screw is tightened. The tourniquet should be screwed up as quickly as possible, that the limb be not charged with blood by obstructing the venous flow, before the arterial current is checked.

To improvise a Tourniquet from a handkerchief, a stone, and a stick. Fold a stone the size of an egg in the middle of a handkerchief, lay it over the main artery, tie the ends of the handkerchief round the limb,

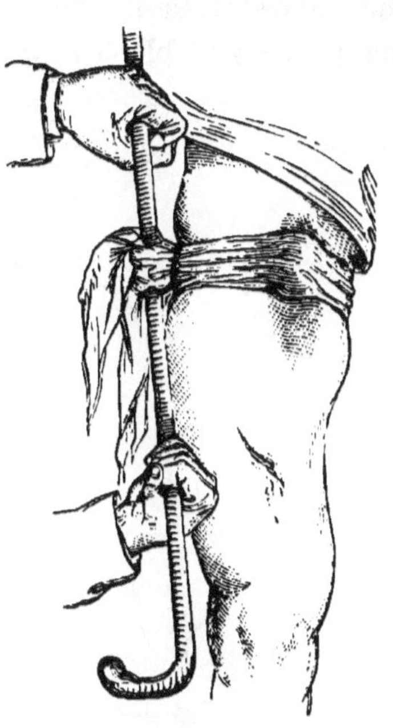

Fig. 112.—An improvised tourniquet.

slip the stick underneath and twist it round, till the tightened handkerchief draws the stone on to the artery and arrests the flow of blood (see fig. 112).

Bloodless Operations.—Esmarch's elastic bandage has been much used of late years to diminish the loss of blood during operations on the limbs.

The apparatus consists of—1. A stout webbed highly elastic roller, about 4 yards long and $2\frac{1}{3}$ inches wide. 2. A clip made of stout india-rubber tubing, 2 feet

long, in each end of which is fixed a wooden plug,
carrying a hook. A better one is Coxeter's elastic
tourniquet.

Before applying the elastic roller the limb is raised
and rubbed towards the trunk in order to diminish
the quantity of blood in it ; the roller is then tightly

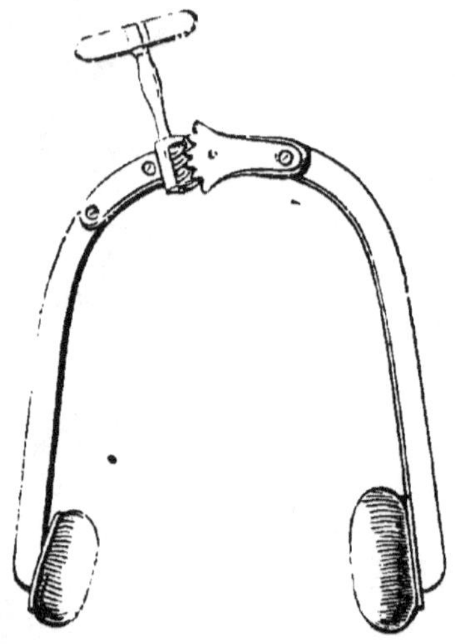

Fig. 113.—The horse-shoe tourniquet.

applied from the extremity upwards in spiral turns,
overlapping each other, and continued for some distance beyond the proposed seat of operation, as, for
example, to the middle third of the thigh for an
amputation of the leg. The elastic tubing is next
drawn tightly round the limb, and the ends hooked
together, close to the highest turn of the roller,
which is then removed, leaving the limb blanched
and exsanguine.

After the operation the main vessels are tied before

loosening the constricting band, which is then gradually relaxed and smaller arteries secured.

In *Signoroni's Horseshoe* tourniquet (fig. 113) the extremities of the shoe can be approximated to each other by a rack screw working a hinge. The ends are furnished with pads, one broad and flat to bear on the limb away from the artery, the other rounded to compress the vessel itself. The tourniquet does not arrest the whole circulation in the limb. It can therefore be applied for a longer time than that of Petit. However, it easily slips out of place, and soon becomes very irksome and painful.

The Abdominal Tourniquet of Lister (fig. 114) is a clamp for compressing the aorta during amputation through the hip joint, and operations where a tourniquet cannot be placed on the limb. It consists of a semicircular bar, with a broad pad to bear on the lumbar vertebræ behind, while in front it holds a long screw-pin carrying a pad. This instrument passes round the left side, and its pad is forced down into the abdomen, 1 inch to the left of the umbilicus, until the aorta is compressed against the spine.

Davy's Lever is a cylinder of hard wood about 18 inches long, which is passed up the rectum far enough for its end to reach the iliac artery. By pressing the lever against the ischium of the sound side, as a fulcrum, the end can be brought to bear on the vessels of the limb through which it is desired to arrest the hæmorrhage during amputation at the hip-joint.

Carte's Tourniquets (fig. 115) are employed to control and diminish the flow of blood through an aneurism. They are intended to be worn for several days, and are fitted with many contrivances for obtaining a continuous pressure on the artery without completely arresting the flow of blood. They are always used in pairs ; in the figure, one presses the external iliac on the pubes, the other the femoral artery. The first is fastened to the body round the hips, the second round the thigh. They are constructed as follows : an arm

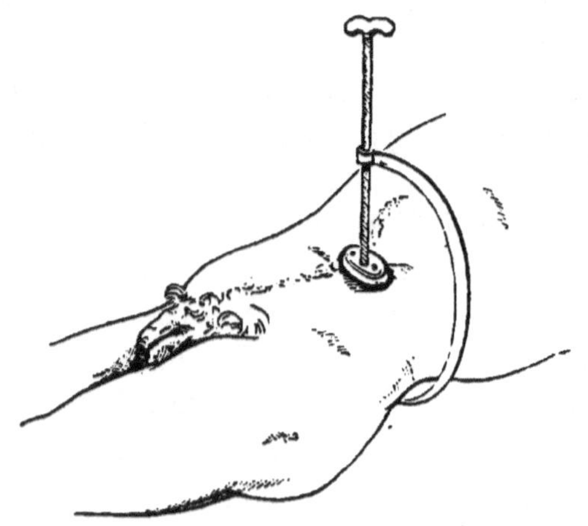

Fig. 114.—Lister's tourniquet for compressing the aorta.

attached to a pad reaches round the limb to the artery, over which it supports a ball and socket joint capable of turning in any direction, but fixed by a screw clamp. This joint has a long screw carrying the compress down to the artery. There is a little play of the screw in the ball of the joint, controlled by india-rubber bands, that the compress may yield slightly before the arterial pulse. In the solidification of an aneurism by this means, the flow of the blood is intended to continue ; hence the current through the vessel should not be completely obstructed by the pressure of the tourniquet, and the elastic bands prevent that pressure from becoming insupportable.

When the tourniquets are applied, the patient must lie on a flat hair mattress, have his limb well washed

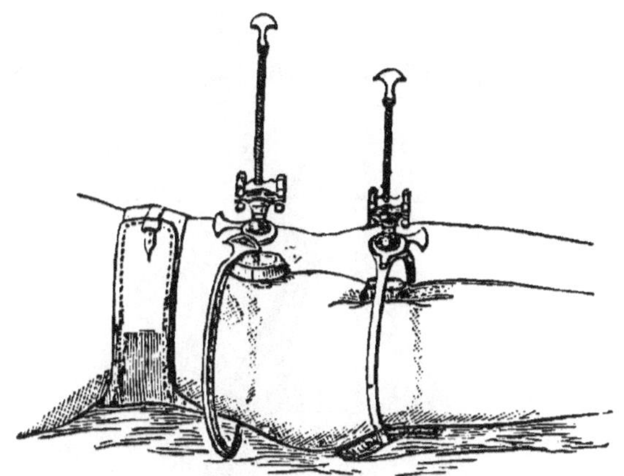

Fig. 115.—Carte's tourniquets for femoral aneurism.

and dried, lightly but evenly bandaged, and somewhat raised. If the thigh is hairy it should be shaved where the pads will press, and dusted with powdered French chalk. The tourniquets are next adjusted, as seen in fig. 115 ; the patient is taught to change the pressure when it grows irksome, by screwing down the second pad, and then releasing the first.

Fumigation.—Mercurial vapour baths are contrived in various ways. The following plan succeeds perfectly well when the whole surface of the body is to be exposed to the vapour (fig. 116).

Apparatus.—A Langston Parker's lamp made by most instrument makers. In this a spirit lamp, holding the required amount of spirit, is enclosed in a cage, on the top of which is a receptacle for the calomel, and a small saucer for water (fig. 117). The flame beneath boils the water and volatilises the calomel. Water is added, because the calomel vapour, when associated with steam, acts more efficiently than with dry air.

The lamp is placed under a high wooden chair, on

Fig. 116.—Mercurial fumigation.

which the patient sits undressed, and round his neck a frame is tied, made of cane hoops, with a woollen cover sewn over them; this falls to the ground and encloses his body in a chamber, where the vapour is confined while being absorbed into the skin. A blanket thrown over the frame completes the prepa-

ration.   The cloak is more expensive, but is a more
effectual cover, when made of mackintosh cloth.

The patient, in four or five minutes, usually breaks
into a violent perspiration, his pulse quickens much,
sometimes even syncope occurs; hence, he should not
be left alone until the bath is over.   This, if the flame

Fig. 117.—Lamp for fumigating.

is strong and the quantity of calomel not very great—
two scruples being a common dose—occupies a quarter
of an hour.   When the bath is over, the patient
should at once get into bed, and lie there a few hours;
then he may rise and be sponged with tepid water.
Moderate but tolerably speedy mercurialisation of the
system is thus induced.

Local Fumigation is employed when the disease
is confined to a few obstinate patches of eruption.
For this purpose an earthenware alembic (fig. 118)
is fitted to the lamp used for general fumigation. The
calomel is thrown into the bottom of the alembic;
the flame plays over the outside, and heating it,
sublimes the calomel, which reaches the mouth of
the alembic and condenses on any part to which it is
applied.

The *throat* may be fumigated by inhaling the vapour as it escapes from this alembic, or by sucking air through the spout of an earthenware teapot in which

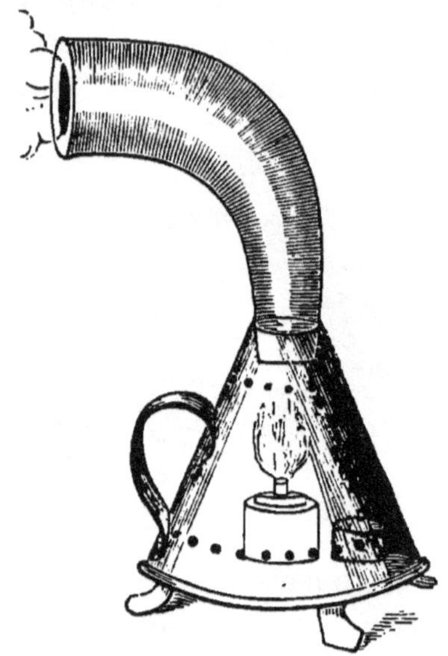

Fig. 118.—Lamp for local fumigation.

the calomel has been placed, heated by a spirit lamp underneath.

The Hot-Air Bath is easily obtained by undressing the patient, putting him to bed on a mattress, and fastening across the bed two or three lengths of cane or stout wire, over which a blanket is next thrown. The patient's body is thus enclosed in a small chamber, the air of which is then heated by putting inside, on an earthenware plate, a spirit lamp, surrounded by a kitchen lemon-grater to protect the bed clothes from its flame. Sheets should be dispensed with while the lamp is alight, lest they catch fire. The temperature of the air should be watched, lest it grow hot enough to scorch, but it must be kept up till the patient breaks into a sharp perspiration, when the lamp may be removed and the patient allowed to cool slowly.

The action of the bath is greatly accelerated by sponging the patient all over as he lies in bed with tepid water, when the air grows warm.

Safety lamps protected with wire gauze, and furnished with a cradle to keep the bed clothes up, are sold at the instrument-makers, and are more free from risk.

The Vapour Bath.—The patient is put to bed as in the hot-air bath, and a few feet of vulcanised india-rubber tubing, fastened to the spout of a tea-kettle on the fire, bring a supply of vapour into the bed.

The vapour bath may precede the hot-air bath, and will greatly quicken the action of the latter.

The Aspirator (fig. 119) consists of a metal-fitted glass syringe, with an accurately adapted piston, and capable of holding from two to four ounces of fluid. The nozzle is connected with two short metal tubes, one at the end, the other at the side, each of which is fitted with a stopcock.

When in use a sharp-pointed hollow needle is fitted to the end tube, while a piece of india-rubber tubing connected with the side tube serves to convey the fluid from the syringe into a vessel. A short length of flexible tubing, interposed between the nozzle and the puncturing needle, protects that from jogs while it is in the wound.

The needle, previously dipped in carbolic oil, is thrust well through the skin, etc., into the cavity to be emptied ; the piston is raised, and the stopcock is then opened, when the fluid will be drawn up into the partial vacuum in the syringe.

When filled, the syringe can be readily emptied, without withdrawing the needle, by turning the stopcock to open by the side tube, through which the fluid is then expelled.

This process is repeated until the cavity is emptied, when the needle is suddenly withdrawn to prevent the entry of air, and the minute puncture left to close of itself.

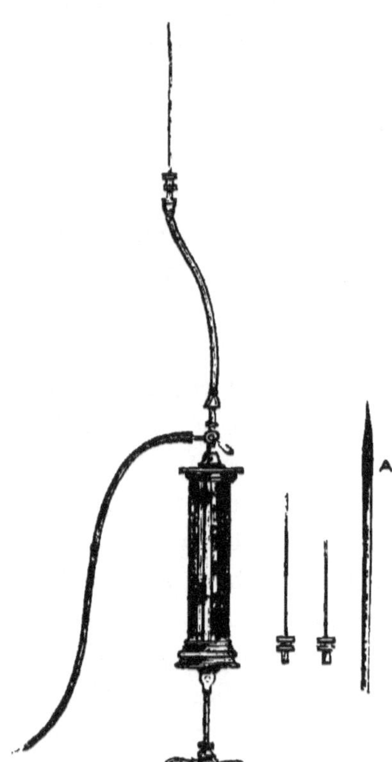

Fig. 119.—The Aspirator.
A. shows the actual thickness of the needle.

Cupping. — *Dry and Wet Cupping.*

Apparatus.—1. A series or nest of exhausting glasses. 2. Boxes of lancets of different sizes for incising the skin, called scarificators.   3. A spirit lamp.

The glasses are 6 oz., 4 oz., 2 oz., and 1 oz. in capacity, of rounded shape, with thick smooth edges (fig. 120).

In *dry cupping* the object is to relieve internal congestion by drawing the blood into the subcutaneous

cellular tissue. The back and loins, where the skin
is tolerably loose, are most suitable places for this
proceeding.

The Operation.

Step 1. Light the spirit lamp, direct the patient to
sit forwards, and lay bare the back ready for the

Fig. 120.—Cupping glasses, lamp, scarificator, and spirit bottle.

glasses, which should be placed on the bed within
reach of the operator's right hand.

Step 2. Rarefy the air in a glass by plunging the
flame into it a few moments, and then quickly clap the
mouth of the glass on the skin; leave it there while a
second and third glass are heated and applied, when
the first should be removed and its vacuum restored
before it is replaced. When the glasses are put on
again, their rims should not lie exactly in the rings
marked on the skin by previous applications, or the
bruises may inflame and slough afterwards at these

parts. The application and removal of the glasses should be done as lightly as possible to prevent all unnecessary pain. The glass is easily lifted by slipping the finger-nail underneath, and thus letting air into the vacuum.

A few repetitions of this incomplete vacuum cause the skin to puff up readily into the glasses, and much blood is thereby attracted into the cellular tissue.

Wet or Bloody Cupping.—When it is desirable to take blood from the body, the skin is first punctured or scarified by the scarificators, half a dozen incisions being made at a blow by as many lancets protruding from a box, when a spring it holds is touched; the glasses are then laid over these incisions, and the necessary amount of blood removed by their exhausting power.

Leeches.—Each leech should draw about 2 drachms of blood, and if the bite is well fomented, another drachm will escape from the wound afterwards.

Before the leeches are applied, the skin should be well washed with warm water, and carefully dried. The leeches should not be taken from their box, but the box inverted over the part, when they will quickly fasten themselves. If the leeches are applied in a dependent position, a soft napkin may be pinned round the box to support them as they grow heavy, and to enable them to suck as long as possible. They should be allowed to drop off; if pulled off they are apt to tear the wound, or leave part of their suckers in it, which causes much irritation afterwards.

The leech is put in a little glass when applied to the

gums or the cervix uteri, and held against the part he
is to suck.

If the leeches do not bite readily the part should
be smeared with blood or warm milk, and the leeches
put into lukewarm water a few minutes ; immersion

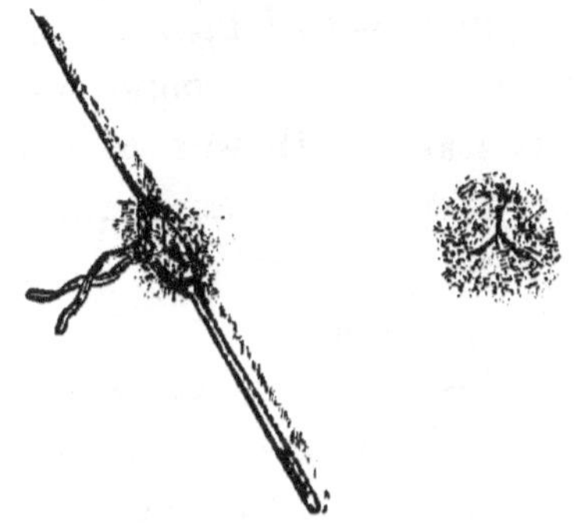

Fig. 121.—A leech-bite ; and a leech-bite secured with a needle.

in small beer is also said to stimulate them to bite.
The bite is tripartite ; three saw-cuts, $\frac{1}{20}$th of an inch
long, radiating from a centre (see fig. 121).

If the bites bleed longer than is desired, they may
be stopped by pinching the skin between the finger
and thumb, wiping the bite thoroughly dry, and filling
it with a little bit of amadou or fine sponge, soaked
in solution of perchloride of iron; a larger piece of
amadou is placed over the first, and the whole compressed with a turn of a bandage or long strip of
plaster. If this fails, a sewing needle may be passed
through the skin beneath the centre of the bite, and
the bleeding surface constricted by twisting a thread
round it under the needle (fig. 121).

Leech-bites should never be left bleeding, especially

in children, for a dangerous amount of blood may be lost from them in a few hours.

Tents are instruments made of some substance that enlarges as it absorbs liquid ; they are employed to dilate apertures of sinuses or natural passages, as the cervix uteri, &c., and are generally short rods 2 to 3 inches long, and $\frac{1}{10}$ to $\frac{1}{4}$ inch in thickness, made of a whalebone stem, wound round with compressed sponge, which is coated with wax to keep it in shape. Slips of gentian root, or of laminaria digitata, which rapidly enlarge as they imbibe moisture, are also employed for this purpose.

Setons are strips of varnished calico, 6 or 8 inches long and $\frac{1}{3}$ of an inch in breadth ; a thread is fastened to each end, which are tied together while the seton is worn. The seton is employed to excite irritation either along the course of a sinus, or in some superficial situation, as the nape of the neck, to relieve congestion of internal parts. In sinuses, a few threads of dentist's silk usually produce the required amount of irritation.

Chassaignac's Drainage-tubes are india-rubber tubes of the calibre of a wheat straw, or larger, of any requisite length, and perforated with holes at short distances. They are carried into the cavity to be drained, by hitching the prong of a forked probe, made for the purpose, through one end of the tube and thrusting it along the sinus, or across the abscess. The skin is then incised over the further end of the sinus to bring the probe out, and the ends of the tube are tied together.

In the Lister's method of treating wounds, drainage-tubes steeped in antiseptic solutions are much employed. Their preparation and use is described (p. 209).

The advantages of these tubes are, the small amount of irritation they provoke, and the ready exit furnished for the matter along their interior.

An Issue is a contrivance for keeping up irritation of the surface. A piece of diachylon plaster the size of a half-crown, with a hole in the centre as large as a pea, is laid over the skin where the issue is to be formed. A bit of potassa fusa is laid in the hole and kept *in situ* by a second plaster, for an hour or till the skin is destroyed under the hole. The plasters are then removed, the wound washed, and a fresh piece of the same size put on, having at its centre a slit $\frac{1}{4}$ inch long, under which a pea is slipped into the sore and covered over by another smaller piece of plaster. The discharge that soon sets up must be washed away twice daily, and the plaster and pea renewed from time to time as they become soiled. Instead of plaster, sometimes a fresh ivy-leaf is laid over the pea, and then two or three thicknesses of old linen. The whole dressing is kept in place, and the wound protected by a small wire gauze armlet, which is buckled on over all.

Trusses for ruptures. These are various in shape, strength of spring, &c.

Whatever variety of truss is employed, care should be taken that the pressure be made in the right direction, and that it be sufficient, but not too great for the strain it has to support.

In reducible hernia the pressure for *inguinal rupture* should be exerted on the inguinal canal and directly backwards (see fig. 122). For *umbilical rupture*, the pressure should be also backwards, and be brought to bear rather on the tendinous margins of the hernial opening than on the aperture itself. In *femoral* rupture the pressure should be directed backwards against the femoral canal (see fig. 123). The pad in all should be large enough to well cover the passage

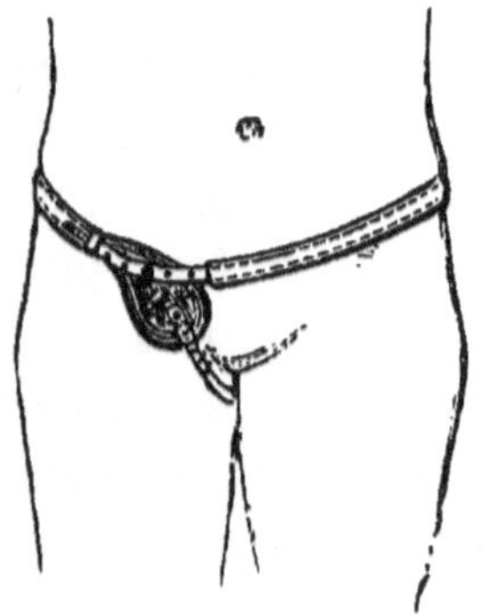

Fig. 122.—Inguinal truss.

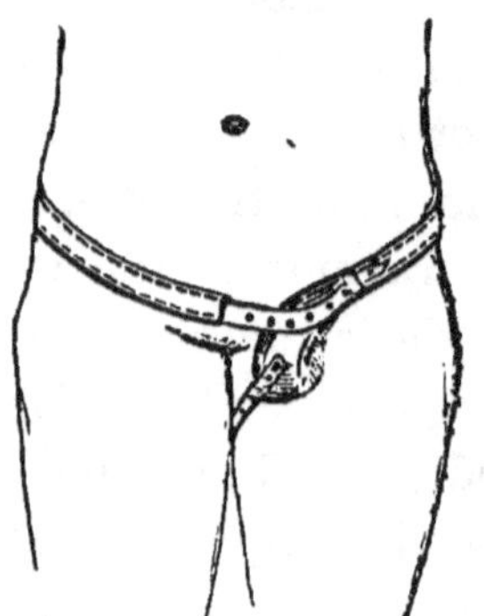

Fig. 123.—Femoral truss.

through which the rupture passes, that the pressure may be exercised upon adjacent structures, rather than directly on the relaxed tissues covering the ring. The efficiency and comfort of a truss much depend on the completeness with which it fulfils these conditions.

The adequacy of a truss should always be tested by directing the patient to separate his legs, lean forward over the back of a chair, and cough or strain deeply. If the truss support the rupture during this exertion it fits satisfactorily.

For *irreducible* hernia, large air-, or spring-padded trusses are made, which prevent further descent of the viscera, but they are exceedingly difficult to fit and often unsatisfactory in use.

In *inguinal hernia* the truss consists of a *pad*, a *spring*, and a *neck* with *guide straps*.

The *pad* is made of various materials ; fine carded wool is among the best, when well stuffed into a properly shaped ovoid leather pad. Mr. Wood recommends the use of a pad made of box-wood and shaped like a horse-shoe, with the inner horn larger and longer than the outer. In most cases a fixed pad is better than a movable one.

The pad should compress the *canal* and be convex if the patient is stout (but in all cases should be as small as will ensure fair compression). A very flaccid belly-wall and a large gap or protrusion require a large surface on the pad. The *spring* should be supple and padded behind to rest on the two sacro-iliac synchondroses, without bearing on the spine. The spring, narrowing as it comes forward, embraces the pelvis ; and opposite the anterior iliac spine inclines down wards, because the hernia is a little lower than the resting-place of the spring behind. When the rupture is almost reached, the spring makes a slight elbow or bend (the neck), that its pressure may be directed against the hernia more fully. *Understraps*, generally not necessary, should be omitted if possible.

In trusses for children, when the testis is not descended, the pad should have a notch at its lower border in which the testis may rest uncompressed.

In the truss for *femoral hernia*, the *spring* bears behind the body and encircles the hips in the same manner as in the inguinal truss, but when opposite the femoral artery it turns abruptly downwards towards the saphenous opening. The *pad* should be somewhat convex, not oval but rounded, and should be placed in the groove of the groin, that most of the pressure exerted on the crural canal may fall exactly between the crural ring above and the saphenous opening below. The *understrap* should be attached to the stud at the lower end of the pad, and pass round the perinæum and fold of the buttock, and be fastened to the spring opposite the great trochanter. It should be made of knitted bandage, that it may be changed and washed frequently.

When measuring a patient for an inguinal truss, the circumference of the body round the hips, between the crest of the hip and the top of the thigh bone (*crista ilii and great trochanter*), should be taken; and this is generally sufficient. When more detail is needed, the distance between the symphysis pubis and the anterior iliac spine, half of which denotes the position of the internal abdominal ring, may be added. This and the inguinal canal are to be supported by the pad of the truss. For a femoral hernia the same measurement round the body should be taken; also the distance of the saphenous opening from the symphysis pubis and from the anterior superior iliac spine. A third measurement from the anterior iliac spine to the symphysis pubis gives the triangle of the groin. These will enable the maker to put the pad at the proper angle with the spring, so that it compresses the saphenous opening, and clears the crest of the pubes.

Every patient should, while he wears a truss, show

himself from time to time to the surgeon to ensure that any defect in his apparatus shall be quickly detected. It is a useful precaution also to keep two trusses at hand, so that if one breaks, the patient may at once apply the other.

Salmon and Ody's truss consists of a spring passing round the hip from a circular pad, which bears on the sacrum to a second oval pad. Both pads are attached to the spring by a ball and socket joint. There is also a slide for shortening or lengthening the spring if desired (fig. 124). This truss is worn

Fig. 124.—Salmon and Ody's truss.

round the *sound* side of the body and reaches beyond the middle line to the hernial opening, with the object of directing the pressure of the spring outwards and backwards, or exactly counter to the course of the hernia inwards.

Umbilical hernia.—Spring trusses are not adapted for restraining umbilical hernia. The support consists of a broad belt fitted to the belly, made in front of elastic webbing, and on each flank, of white jean. Behind, the belt is fastened by straps and buckles, or by lacing, the better plan. In the centre, the elastic part carries a nearly flat air-cushion, measuring about 3 inches transversely and $2\frac{1}{2}$ vertically. This cushion is placed against the aperture of the belly, and presses back the protrusion. The size of the pad varies with the size of the hernia, but it should always largely exceed the extent of the gap in the abdominal wall. The pad, when the apparatus is used for an infant, should not be too prominent, as it is then more difficult to keep in place, and also, by pressing into the aperture, hinders it from closing. The pad for an infant is best made of a disc of ivory, $1\frac{1}{2}$ inch broad and $\frac{1}{2}$ an inch to 1 inch thick, stitched in a little case in the centre of the girdle. Mr. Wood recommends the employment of an oval cup-shaped pad, made of vulcanized india-rubber, by which the pressure is brought to bear on the tendinous margins of the hernial opening in such a way as to tend to close them towards the median line. The quantity of elastic tissue should be much less in the infant's belt than in those for adults, that the belt may be frequently washed. The difficulty of keeping the apparatus in place is lessened by attaching two bands to the upper border, to pass over the shoulders and cross behind before fastening to the belt, like braces. Two similar ones may be fastened to the lower border and carried under the thighs. These bands should be of soft webbing, and several pairs kept in store, that they may be frequently changed and washed.

Cauteries.

Cautery irons.—These are masses of iron of different shapes; some pointed, others rounded like buttons, &c., set in a stem a foot long, fixed in a thick wooden handle. They are heated in a charcoal brazier or common fire to bright redness if required to destroy deeply, but short of redness if intended only to scorch the surface.

As these irons are inconvenient for many cases from their bulk, and yet soon lose their heat if made small, other cauteries have been devised to which the heat can be quickly renewed, such as Nelaton's or Bruce's gas cauteries, or Paquelin's benzole cautery.

Galvanic cautery.—The instrument consists of a platinum wire, made to glow by passing through it a powerful galvanic current. The wire should be thick (about $\frac{1}{12}$ of an inch), and all the other conducting surfaces sufficiently large to offer no impediment to the current where heat is not desired.

The main advantage of a galvanic cautery is that the wire can be passed while cold exactly where it is required, and then heated when it is in place. Again, by this means, an intense heat can be applied to a very limited area, and more quickly renewed than by any other plan, for the wire, even when plunged in the tissue, is never far below a red heat.

The platinum wire has been arranged in several forms, to suit different requirements, the most useful being the instruments of Middeldorpf.

Paquelin's cautery has superseded other cauteries for most purposes from its handiness. It consists of a platinum point made hollow, and into which a stream of benzole vapour mixed with air is driven. The ignition of the combustible mixture in the interior of the platinum causes that metal to grow hot even to whiteness, and the heat is easily maintained. The accessories to the platinum cautery point are a Higginson's pump and wash-bottle connected with the cautery by an elastic tube. Through them an assistant keeps up a constant stream of air by the

Higginson's pump, which, forced through the wash-bottle of benzole, is charged with the vapour of that fluid. A spirit-lamp is added to heat the platinum red hot before the benzole vapour is pumped through it, otherwise the apparatus is apt to miss fire.

Of *chemical caustics* a host exist ; those most commonly employed are :—nitrate of silver, solid, or in saturated solutions (2 drachms to the oz. of water, &c.) ; fuming nitric acid ; solution of nitrate of mercury in nitric acid ; oil of vitriol made into a paste with powdered charcoal ; chloride of zinc mixed with dry starch, then rolled into cakes and cut in slices ; Vienna paste—that is, equal parts of potassa fusa and quicklime worked into a paste with spirits of wine ; potassa fusa itself ; solution of chromic acid. Some surgeons prefer one, some another ; as a rule, the liquid caustics are employed where the surface to be destroyed is uneven and spongy, and solid caustics where the surface is smooth, and a long-continued action is desired.

Vesicants and irritants.—Of the commonest are *mustard poultices*, made by mixing mustard flour in a basin with luke-warm water—*i.e.*, about 100° F.—to a paste and spreading it on muslin, which is again folded over the exposed surface of the mustard. Boiling water and vinegar should not be used, for they lessen the pungency of the poultice. If the full effect be desired, the poultice should remain on the skin fifteen or twenty minutes. If only slight reddening is wanted, the mustard flour should be diluted with its bulk of linseed meal before mixing it with water.

A stronger vesicant is *Corrigan's hammer*, a button of polished steel with a flat surface, fixed to a handle.

When used it should be plunged for a couple of minutes in boiling water, or heated over a spirit lamp, but care must be taken not to overheat it, or it will bring the cuticle away with it. It is pressed on the skin for ten or fifteen seconds ; this is sufficient to cause reddening and vesication.

Blisters are raised by the emplastrum lyttæ, Bullen's liquor epispasticus, or the *pâte épispastique*. This last is milder in its effect than the two preceding preparations of Spanish fly. A solution of iodine and iodide of potash, in three times their bulk of spirit of wine, also produces a blister when laid on freely.

Poultices are made of linseed meal, bread, or starch, and are means for applying warmth and moisture without absolutely wetting. Of the three kinds the bread poultices sodden the parts to which they are applied the most, and starch the least.

Before making a poultice, all the materials should be at hand and thoroughly warmed before a good fire. They are—boiling water, a broad knife or spatula, soft old linen or muslin, oiled silk, tapes, strapping plaster, bandages, a piece of old blanket, flannel or cotton wadding, safety pins, or needle and thread.

The linen on which the poultice is to be spread should be cut of the intended size, and when for use about the neck or shoulder should have some tapes sewn on to it to tie it on to the body. The oiled silk should be large enough to cover the poultice next which it is laid, to keep in the moisture. The flannel or wadding is used to wrap over and keep in the heat of the poultice ; the strapping or bandage to fix everything *in situ* as required.

When poultices are continued for a long time, their surfaces should be spread with lard before application ; this protects the skin somewhat from the irritation that arises ; also, when the poultice is to be laid between folds of skin or on hairy situations, as the buttocks and perinæum, it is better to cover the poultice with a thin cambric handkerchief lest some of the meal stick to the parts.

The Linseed Poultice is made as follows : pour boiling water into a well-heated basin till the basin is half full, then scatter meal with the left hand on the water while that is kept continually stirred with a broad knife, adding more and more meal until the mass becomes quite soft and gelatinous, but too stiff to cling to the knife ; then turn it out on to the linen, also well heated at the fire, and spread it in a layer about $\frac{1}{8}$ an inch thick, turn up the edge of the linen for $\frac{1}{2}$ an inch all round. Carry the poultice at once to the patient ; (if it has to be carried far, the poultice should be laid between two very hot plates). Apply it to the part to be poulticed, lay on the oiled silk, and cover that with the hot flannel or cotton wadding, and fasten these in place with pins or a stitch. Wadding is put where the part is irregular, as the neck or axilla ; unless the wadding is well placed and the poultice is fastened by strings, it will soon fall into a narrow band, leaving exposed the part that it should warm and moisten.

The Bread Poultice is made as follows : the materials being all at hand, as detailed in the directions for making a linseed poultice, crumble the inside of a moderately stale loaf until about half a pint or a pint of crumbs are prepared ; then pour boiling water into a basin, and throw in crumbs gradually in the same manner as the linseed meal, until a soft porous mass is prepared. The remaining steps are the same as those for making the linseed poultice.

The poultice can be made to hold more water if it is turned into a saucepan after mixing, and a little more water added while it simmers for half an hour at a slow fire. Any superfluous water must be drained off, and the poultice covered with muslin when it is made in this way.

A cold Bread Poultice Higginbottom directs to be made as follows : "Take a penny loaf, remove the crust, put the soft part in a pan with a pint of cold water ; let it boil gently for an hour, stirring constantly till it is of a medium consistency; when cold, spread on linen half an inch think. Put the rest in a basin for use, covering it to keep it damp. Use *no* grease or oil."

The Starch Poultice is made as follows : rub a little starch in a basin with cold water till it has the consistence of cream, then mix in *boiling* water till the starch is a thick jelly, and spread it on the linen while hot. Starch poultices retain their heat a long time, but yield very little moisture to the part. They are chiefly used as emollients to inflamed affections of the skin, &c.

A substitute for a poultice consists of a sheet of *spongiopiline*, well soaked in boiling water, and applied under oiled silk : or several folds of lint soaked in boiling water, and then gently rung out, may be used in the same way.

Hot fomentations are a means for applying heat when moisture is not desired. A ready mode is to take a piece of blanket or thick flannel, soak it in boiling water and dry it by wringing in a folded towel, and then wrap it over the part to be fomented with a piece of oiled silk or a hot dry flannel over it. If the fomenting flannel be rolled up and held upright while boiling water is poured from a tea-kettle into its interior, it can be made thoroughly hot and needs no wringing afterwards. The nurses' hands, which may be further protected by a pair of hedger's gloves, are thus saved from scalding. Laudanum, turpentine, and other applications are sprinkled over the flannel, when soothing or counter-irritating effects are required in addition to the warmth. A bag of bran makes a light warm fomentation if heated in a steam kitchen, or steamer for boiling potatoes.

When absolutely *dry heat* is desired, chamomile flowers, salt, bran, sand, or bricks and tiles, may be heated in an oven, and poured into hot flannel bags.

Dry heat is also very agreeably obtained by filling *india-rubber bags* and *cushions* with hot water ; they are rather heavy, but retain their heat many hours.

Antiseptic Treatment of wounds.

Lister's method of dressing wounds.

Carbolic Acid.—The properties which concern the surgeon may be briefly recapitulated as follows.

The acid is highly volatile, but while present in sufficient quantity indefinitely postpones the putrefaction of organic fluids. It is soluble in different degrees in water, alcohol, ether, glycerine, fixed oils, gutta percha, india-rubber, resin, &c. Its varying affinity for these substances enables the surgeon to modify the application of the acid in various ways ; these modifications are necessary fully to utilise its properties.

Water dissolves the crystallised acid but sparingly, 1 part in 20 being a concentrated solution, and allows it to escape readily ; the aqueous solution is therefore useful when the effects of the acid are required copiously, but only temporarily. Glycerine and the fixed oils dissolve a far greater amount of the acid than water does, and the solutions, as regards tenacity, hold an intermediate place between the aqueous solution and the resinous mixture. This latter substance, while forming a mild preparation, has likewise the property, owing to its strong affinity for the acid, of storing it up for a considerable time.

Carbolic acid stimulates raw surfaces, and when concentrated, even destroys animal tissues. It is a local anæsthetic ; on its application in solution of moderate strength, wounds lose their sensibility after the first smarting has passed off. When absorbed in large quantities, the acid first induces vomiting and diarrhœa, and after large doses a peculiar kind of delirium often ensues, followed by temporary paralysis of sense and motion ; fatal results have followed its internal administration.

It is rapidly absorbed into the blood from wounded surfaces, as well as through the unbroken skin, and is thence discharged from the body by the lungs and kidneys. In most cases the urine of a patient with a wound dressed with carbolic acid, although of normal colour when passed, assumes a dark greenish-brown hue after a few hours' exposure to the air and light.

The following are the preparations and materials employed. *Watery solutions* of 1 part of acid to 20 of water are used for purifying the epidermis of a part about to be operated on; for cleansing dirty instruments and sponges, and also for washing accidental wounds, so as to destroy any septic organisms that may have been introduced into them.

A solution of 1 part of acid to 40 of water is employed for the spray, for squeezing out the sponges, for wetting the hands of the surgeon and assistant, and for remoistening instruments laid aside temporarily during an operation; for washing the surface of wounds while the dressings are changed, and for soaking the linen rag used as a guard to cover the wound whenever the spray is intermitted.*

Oily solutions of the strength of 1 part of acid to either 10 or 20 parts of olive oil are employed for various purposes, among others, for moistening the sharp cannula of the aspirator previously to using it, for soaking strips of lint or gauze to be used as drains, &c. These solutions are also applied on lint to the interior of wounds where it is desirable to have the constant active operation of an antiseptic, which admits of being frequently changed.

Antiseptic gauze.—This consists of a light cotton

---

* A solution of 1 part of the acid to 5 parts of spirit may be injected through an elastic catheter connected with the syringe by means of a piece of india-rubber tubing, into the recesses of the wound caused by a bad compound fracture, when first seen some hours after the accident.

material prepared by being charged, in a proportion nearly equal to its own weight, with a compound of 1 part of crystallised carbolic acid, 5 parts of common resin, and 7 parts of solid paraffin.

For a dressing, this material is folded in eight layers, of sufficient size to cover an area extending usually for 8 inches in every direction from the wound. Between the two outermost layers is inserted a sheet of very thin macintosh cloth, of the description commonly known as hat lining; the object of this impermeable layer is to diffuse the discharges evenly throughout the folds of the dressing, in order to prevent the possibility of their soaking through the gauze at any one point, and so exhausting the carbolic acid at that spot, without utilising the rest of the dressing.

The protective consists of ordinary oiled silk, which is first coated with a layer of copal varnish, and when dry is brushed over with a mixture of dextrine and starch. This protective, first moistened in 1 to 40 solution, is placed over the wound beneath the dressing, to protect its surface from immediate contact with the carbolic acid contained in the gauze. It is used only in cases where it is desirable to procure speedy healing.

Vulcanized india-rubber drainage tubes are used, varying in size from a crowquill to that of the little finger; at various intervals holes are cut in them, equal in diameter to half that of the tube. One or more tubes, according to the extent of the wound, should be inserted to the required depth, and their ends be cut level with the surface. They are kept in

P

position by two pieces of silk passed through opposite sides of the orifice and knotted at the extremity furthest from the wound.

As the cavity of the wound gradually fills, the granulations tend to push the tube out, and it will therefore require to be shortened from time to time; but this should not be done too hastily. Its use should be continued in abscesses until the cavity is completely closed, and in wounds so long as even trifling oozing exists. The drainage tubes should be kept immersed in 1 to 20 watery solution.

Either *carbolised silk* or *metallic sutures* may be employed in operations under the spray. The former are prepared by immersing a reel of silk in a mixture of melted bees-wax, with about one-tenth part of carbolic acid, the thread being drawn through a dry cloth as it leaves the liquid, to remove the superfluous wax.

Carbolised catgut is used for ligaturing vessels when torsion is not advisable; both ends of the ligature are cut short. The catgut should be kept immersed in an oily solution (1 part to 5).*

The *spray producer* commonly employed consists of a modification of Siegle's steam spray producer.† A 1 to 20 watery solution of the acid is placed in the glass bottle which forms part of the apparatus, and this, mixing with the steam from the boiler, makes the spray of about the strength of 1 to 30. By this means an atmosphere charged with carbolic acid is

---

* For preparation, see Mr. Lister's article on "Amputation," in 2nd edition of "Holme's Surgery," vol. v., p. 622.

† To be obtained of Young, Edinburgh, or Mayer and Meltzer, or of Allen, or of Marr, or Matthews, in London.

maintained in the neighbourhood of a wound during an operation or the changing of dressings.

The following are the materials required for dressing wounds with carbolic acid :—

1. Watery solutions (1 part in 20 and 1 part in 40). 2. Oily solutions (1 part in 10 and 1 part in 20). 3. Old linen—lint. 4. Prepared catgut ligatures. 5. Drainage tubes of various sizes. 6. Carbolised silk or metallic sutures. 7. Oiled silk protective. 8. Antiseptic gauze. 9. Hat lining (*i.e.*, thin macintosh cloth). 10. Loose pieces of gauze to be used for packing. 11. Gauze bandages. 12. Safety, or nursery, pins. 13. Spray producer. 14. Syringes. 15. A gum elastic catheter. 16. Piece of india-rubber tubing. 17. Sponges soaked in 1 to 40 lotion. 18. Elastic webbing bandages of various widths.

To dress a recent wound.

1. All hæmorrhage having ceased, wash the part, and syringe out the wound thoroughly with 1 to 20 lotion under the spray ; and then cover the wound with a guard soaked in 1 to 40 lotion.

2. Prepare the dressing, if this has not been previously done, by folding gauze into eight layers of a size sufficient to envelope the part ; place a sheet of hat lining of equal dimensions between the two outermost layers.

3. Cut a piece of protective a little larger than the surface of the wound ; then remove the guard under the spray, and apply the protective, moistened in 1 to 40 lotion, over the edges of the wound.

4. Place two or three folds of gauze wrung out of 1 to 40 solution over the protective, large enough

to overlap it, and apply the dressing, wrapping it carefully over the part, and taking care that the centre of the gauze be placed immediately over the wound, and that the spray be steadily maintained until the part is well covered in.

5. Fix the dressing securely by carrying a gauze roller evenly and firmly over and around the mass, taking care well to cover in the edges. Fasten down the dressing by an elastic bandage carried along its margins to ensure that the dressing shall fit closely to the skin during the patient's movements.

6. Finally secure the bandage with safety pins so inserted as to prevent the possibility of the dressing shifting at any part. Care must be taken not to thrust the pins through the sheet of hat lining, or its efficacy will be destroyed.

Should the surface of the part to be enveloped by the dressing be uneven, as in the neighbourhood of a large joint, or should the discharge be likely to prove very abundant, it is a good plan to place loosely crumpled-up pieces of gauze inside the dressing before closing it; these serve to adapt it closely to the part.

In cases of *compound fracture*, the wound is first dressed as above described, while an assistant maintains the limb in good position; padded splints are then applied over all.

In changing a dressing, care should be taken, by placing the hand over the seat of the wound, to prevent the gauze from being raised from the surface before the spray can be directed over the wound.

All dressings should as a general rule, be changed within 24 hours at furthest from the time of being first applied ; but during the subsequent progress of the case, the gauze may be left undisturbed for periods varying from two days to a week, provided that no discharge shows beyond the edge of the folded dressing.

In situations where there is not a sufficient surface of skin available for the gauze to extend beyond the vicinity of the wound, for example, about the pubes after herniotomy, the deficiency in size may be compensated by increased thickness of the dressing, the gauze being used in 16 or even 32 layers instead of 8.

To perform an ordinary operation, such as an amputation or the excision of a tumour.

1. Purify the skin with 1 to 20 watery solution.

2. Make the necessary incisions under the spray, the instruments, sponges, and the hands of the operator having been previously purified by dipping into a watery solution of the acid, and the neighbourhood of the part to be operated upon being guarded by towels similarly soaked.

3. Secure all vessels with carbolised catgut, as fine as is consistent with a due amount of strength.

4. Insert drainage tubes, varying in size and number with the size of the wound ; remembering that more oozing takes place during the first 24 hours from a wound to which carbolic acid has been applied than from one which has not been so treated.

5. Bring the edges of the wound into very accurate apposition by numerous sutures, leaving only an interval at one end for the extremities of the drainage tubes. This is most securely achieved by Lister's button suture, described on p. 216.

6. Apply protective, moist gauze, and after packing inequalities with pieces of loose gauze, adjust the main dressing as before.

To open an abscess.

1. Cleanse the surface of the part with 1 to 20 lotion, and shave it if necessary.

2. Make the required incision under the spray, with a knife previously dipped in 1 to 20 solution.

3. Having pressed out the pus, insert a drainage tube, or a strip of lint soaked in 1 to 10 oily solution.

4. Next place over the incision a small piece of gauze wetted in the watery solution, and over all apply a dressing as above directed.

In cases of *chronic effusion of fluid in the bursa patellæ*, or of *small abscesses* in situations where a scar is undesirable, the fluid may be evacuted under the spray by a puncture with a tenotomy knife ; a small drainage tube is then inserted, and a pad of gauze applied and retained by a bandage.

In the case of *ischio-rectal abscess*, the parts should be first well washed in 1 to 20 solution, and the abscess opened under the spray ; a pad of lint soaked in a 1 to 10 solution of the acid in glycerine or oil, should then be applied and retained in place by a T bandage, and this dressing should be changed every five or six hours.

The patient should be directed before defæcation, to draw the pad and bandage to one side, at the same time keeping the wound covered. The parts should be well cleansed with the solution, before the pad is readjusted.

Chloride of zinc in aqueous solution of 40 grains to the ounce is a powerful antiseptic. A single application of it to a recent wound, though without producing a visible slough, will prevent the occurrence of putrefaction in the cut surface during a period of from 48 to 72 hours. Hence it is applied to operation wounds in situations which render the subsequent avoidance of putrefaction impossible, as for example about the jaws, in the vicinity of the anus, and in the neighbourhood of sinuses due to caries. It should not be applied to surfaces which are expected to unite by the first intention.

The *sharp spoon* is an instrument recommended by Professor Volkmann, of Halle, for scraping carious bone. He also employs it for clearing out the pyogenic membrane of putrid abscesses and sinuses, and removing all granulations around the diseased bones after excision.

Mr. Lister has of late employed a modification of the quill suture, which he has named the Button Suture (see fig. 125). By this method a wire is carried across the bottom of the wound so as to bring a large surface of the flaps into close adaptation ; the wire also maintains the parts in apposition, and prevents them from being lifted asunder by serous effusion, swelling, or by unavoidable movements when the dressings are changed. To release the tension if the swelling of the parts afterwards need it, the wires are easily detached from the leaden plates, and again made fast. So, likewise, the stout wire stitches of relaxation (*b* in fig. 125), can be loosened when needed.

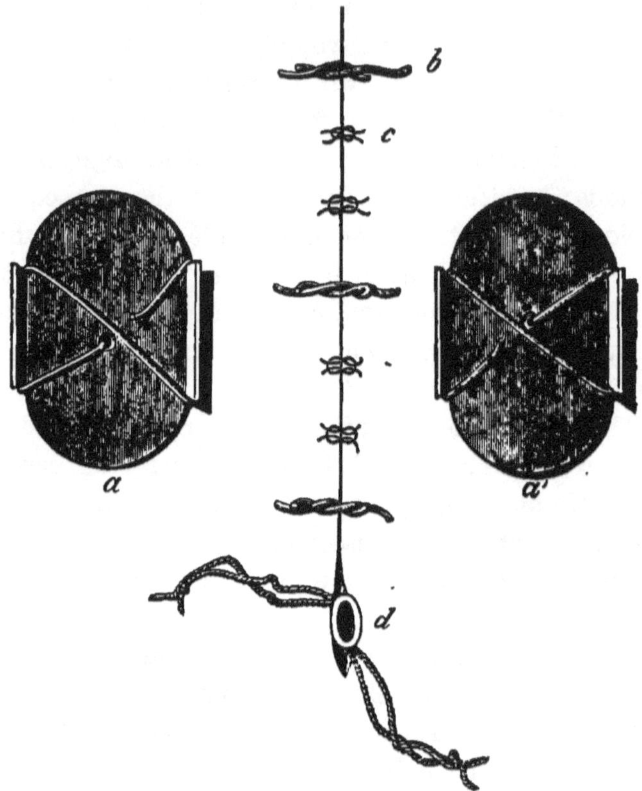

Fig. 125.—Lister's Button suture.

a a′. Lead buttons united by stout silver wire passed through the substance of the flaps.

b. Stout silver wire stitches of relaxation, passed at some little distance from edges of flaps.

c. Fine wire, or silk or horse-hair stitches of coaptation, passed as close as possible to the edges of the flaps.

d. Mouth of the drainage tube, cut level with the surface of the skin, and with two pieces of silk attached to the orifice to prevent it from slipping into the wound.

Boracic acid is an extremely bland and unirritating antiseptic, but, owing to its non-volatility, is applicable only for the treatment of superficial wounds. The crystals, which are very soft and unctuous, are much more soluble in hot than in cold water.

The acid is employed in three forms :—

1. As a cold saturated aqueous solution.

2. As boracic lint; this is prepared by immersion in a boiling saturated solution, whereby the lint absorbs nearly its own weight of the crystals.

3. As boracic ointment of different strengths.*

To dress ulcers of the leg.

1. First cleanse the sore with solution of chloride of zinc (40 grains to 1 ounce), or with iodoform, and well wash the surrounding skin with carbolic lotion (1 to 20).

2. When the discharge has become quite sweet,

---

* *Martindales's Boracic Ointments.*

No. 1.		No. 2.	
Paraffin . . . 5 parts		Paraffin . . . 5 parts	
Vaselin . . . 5 ,,		Vaselin . . . 10 ,,	
Boracic Acid in fine		Boracic Acid in fine	
powder . . . 2 ,,		powder . . . 2 ,,	

No. 1 is for spreading on lint.

No. 2. is softer, for anointing the skin directly.

In preparing the ointment it is necessary that the boracic acid should be in a state of fine powder, and further that the paraffin and vaselin be thoroughly rubbed together while the ointment is compounding. If this is not done, the paraffin will remain lumpy and the ointment will want smoothness.

apply a piece of protective, rather larger than the sore, moistened in boracic lotion.

3. Place over this a fold or two of boracic lint, soaked in the lotion, and large enough to extend for an inch or more over the edge of the protective.

4. Finally, retain the dressing with an ordinary bandage.

This may be left undisturbed for a period varying from two to five days, according to the amount of discharge. It is often more convenient to purify the sore by means of the boracic lint soaked in the boracic lotion changed every few hours and without the intervention of the protective.

The lint soaked in boracic lotion may also be used in the same way as an ordinary water dressing to foul ulcers, or to deep burns with foul sloughs. It then requires changing once a day or oftener.

The ointment may be employed spread on thin linen, to dress superficial ulcers—no protective is then used.

Salicylic acid, recommended by Professor Thiersch, of Leipzig, has come into use as an efficient antiseptic.*

This acid is inodorous, non-volatile, and less irritating than carbolic acid ; it is soluble at the ordinary temperature of the air in the proportion of 1 part to 300 of distilled water.

The following are the preparations employed :—

1. An *aqueous solution* (1 part of acid to 300), used for washing and irrigating wounds, and as spray.

* See the *London Medical Record* of May 26th and June 2nd, 1875, pp. 317 and 334.

Salicylic spray causes sneezing and coughing, hence the carbolic spray is often substituted for it.

2. *A dressing*, consisting of cotton wool charged with the acid in the respective proportions of 3 and of 10 per 100.* The wool of the latter strength is usually coloured with carmine to distinguish it from the weaker preparation.

Iodoform is now much employed instead of carbolic and salicylic acids. It is used as powdered crystals or as a solution in oil of eucalyptus, or in ether; or suspended on cotton wool by soaking the wool in the etherial solution and afterwards drying it. Wounds are dusted freely with the powder or painted with the solution, while the wool is used to cover them and receive their discharges.

Irrigation.—The continual flow of ice-cold water is used to prevent inflammation of certain wounds. In using cold, it is particularly necessary that the temperature of the water remain steady, for alterations of temperature cause alterations in the capacity of the blood vessels, and promote congestion rather than diminish it; hence irrigation, badly attended to, becomes an evil instead of a benefit. The simplest way (see fig. 126) of contriving irrigation is to lay the limb in an easy position on pillows, protected by a sheet of india-rubber cloth, weighted at one corner to draw the cloth into a channel, down which the water trickles into a receiver under the bed; over the limb a jar, wrapped in blanket, is suspended. This is filled with water from time to time, and kept charged with

---

* The prepared wool may be obtained in London of Messrs. Krohne and Sesemann, Duke Street, Manchester Square.

lumps of ice. A syphon is made by a few feet of
fine india-rubber tubing reaching from the bottom of
the jar to the wound, the escape of water through
the tube being moderated by drawing the end more

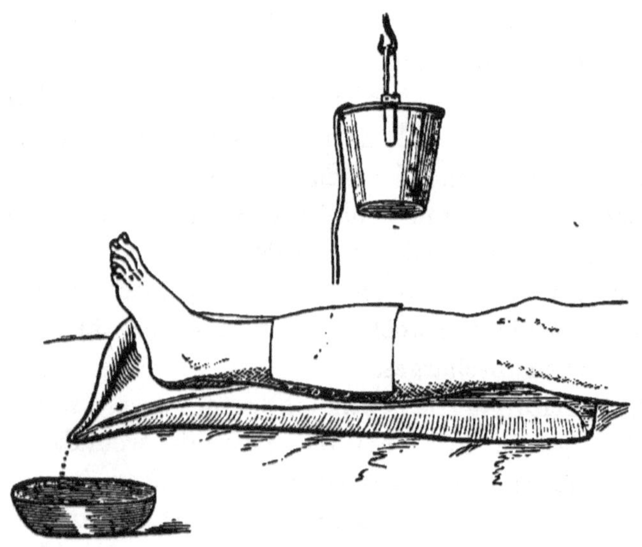

Fig. 126.—Irrigating a wound.

or less tightly through a bit of cleft stick. It is
sufficient that the wound should be kept constantly
and thoroughly wetted ; more than that is waste of
cooling power.

A spiral coil of india-rubber tubing disposed round
the part to be cooled is a more rapid and efficient
method of refrigeration ; the coils, made large enough
to fit loosely, should be held together by an interlacing
narrow tape.   When in use, one end weighted to make
it sink is put into a vessel of ice-cold water. placed two
or three feet above the patient, and the other end,

provided with a small tap to regulate the flow, is dropped into a receiver on the floor. If the receivers are wooden buckets, well wrapped in blanket, the occasional addition of a lump of ice keeps the temperature of the water sufficiently low, and by changing the position of the vessels as the upper one empties and the lower one fills the current is made to pass backwards and forwards as long as required.

Leiter's tubes are fine gaspipe metal tubes laid up in coils, and the coils bent into various shapes, so that any part of the body may be fitted with a coil through which water flows continuously, and thus a constant supply of heat or cold obtained as may be desired.

Esmarch's Irrigator.—This is a simple contrivance

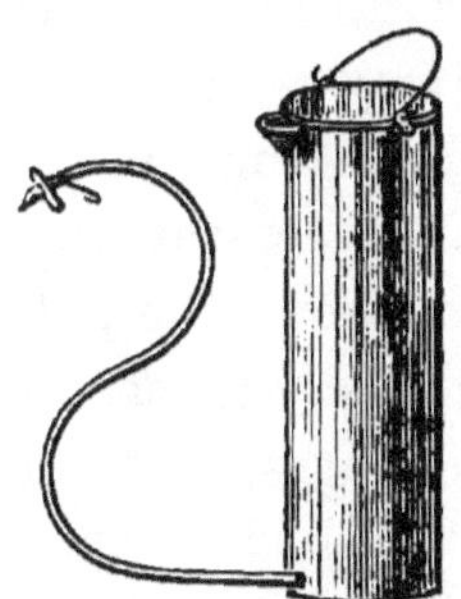

Fig. 127.—Esmarch's Irrigator.

for washing out wounds and sinuses with a stream of water. It consists of a tall can of block tin (see fig. 127), with an orifice at the lower end, to which a couple of feet of india-rubber tubing are attached. The tube is fitted with an ivory nozzle and a hook, so that when the stream is not wanted the flow of water is stopped by hanging the nozzle on the upper edge of the can. The stream can be made more or less forcible by raising or lowering the can above the wound.

The Administration of Chloroform.—In administering chloroform the main points to be borne in mind are—1. That if the patient is fit to undergo an operation at all he may inhale chloroform. 2. The patient should be fasting; at least four hours should pass between the last meal and the operation; especially must all stimulants be avoided; this is the most effectual preventive of sickness. 3. He should be in an easy position, clad in a loose but warm night-dress, which does not interfere with ordinary or with artificial respiration, should that be suddenly required. 4. The patient must never inhale more than 4 per cent. of chloroform vapour in the air he respires; on the other hand, the vapour may circulate in the blood without harm for an indefinite time, provided it never passes beyond a certain concentration. 5. Chloroform is a sedative and depressant; the pulse gives the earliest indication of syncope; it should therefore be constantly under the finger, but the respiration should also be carefully watched during the whole administration. The pulse often falls suddenly at the first flow of blood in an operation, and syncope may occur. In such condition inhalation should be immediately stopped. Again, when the patient is deeply narcotised, the jaw may gape and the tongue sink back till it closes the glottis. From this cause respiration sometimes ceases, and danger quickly arises if the chin is not drawn up to raise the epiglottis. In beginning to inhale, the quantity of vapour should be small, and gradually increased. The patient must be cautioned not to talk, to avoid the irritation and coughing chloroform sometimes excites while he is speaking. He should also shut his eyes lest the vapour make them smart. After inhalation has been continued for a few minutes, the patient is often quiet and inattentive, though easily roused by pain. His condition at this stage should be tested by asking him to give his hand, or by pinching him gently; if no notice be taken of these stimuli, the conjunctiva should be touched, and the amount of winking thus excited will enable the chloroformist to judge if the patient will resist when the knife is applied. Patients vary much in the time passed before recovering consciousness; if they remain soundly asleep, breathing freely and with good pulse, it is better to avoid rousing or moving them until they wake spontaneously; such patients suffer less from confusion and vomiting than those who are quickly alive to what is going on around them.

Signs of Danger.—Sudden failure or irregularity of the pulse, with pallor, or dilatation of the pupil, or any alteration of the respiratory movements, are of great importance. If any such changes occur, the chloroform must be at once removed, a free supply of fresh air ensured, by suddenly compressing the chest walls, the tongue drawn well forward with a forceps. If the breathing do not quickly improve, artificial respiration must be set up immediately (see p. 228) and continued for at least an hour before recovery is despaired of. Stertorous breathing is not alarming unless accompanied by feeble pulse, shallow respiration, and dilatation of the pupils ; with these it becomes a sign of a comatose condition.

As subordinate adjuvants for faintness the following are useful : moistening the tongue and lips with brandy from time to time, or letting the patient sip a small quantity from the spout of a feeding cup.   In complete syncope, the continuous galvanic current to the epigastrium, or a hot iron or scalding water to the præcordia may be employed, but *nothing should ever interfere with the maintenance of artificial respiration, which is of far greater efficacy in restoring suspended animation than anything else.*

The inhalation of four or five drops of nitrite of amyle from a handkerchief is a powerful restorative in syncope.

Elevation of the trunk and lower extremities, so that the head hangs perpendicular for a few seconds, has rapidly revived patients from severe syncope.

The best method of administering chloroform is by Clover's inhaler from a bag containing a definite mixture of air and chloroform ; but it may safely be given on a handkerchief, or in various ways, if the administrator is careful to watch the respiration and pulse, and to guard against the patient's taking, by a sudden deep inspiration, too large a dose of vapour at once.   Exact measurement of the quantity of liquid poured on the handkerchief at a time is misleading, as it is no index of the concentration of the air respired by the patient. Of far greater consequence is it to insure a free supply of atmospheric air, by keeping the evaporating surface a few inches from the mouth and nostrils.

The Administration of Ether.—The rules which
have been already laid down to be observed when
giving chloroform apply to the administration of ether.
Ether depresses less than chloroform, hence is less
liable to produce syncope, but it affects the air passages and therefore may cause considerable bronchial
irritation, and the copious secretion of mucus. The
inflammability of ether vapour renders it unsuitable for
operations about the face where the actual cautery is
used. Again, during operations in the mouth, greater
concentration of the vapour of ether than of chloroform
vapour is required to preserve anesthesia ; this necessity makes the latter preferable, because in such cases
the patient of course has his mouth widely open the
whole time the surgeon is employed.

Ether may be safely inhaled from a towel folded
into a cone, and fastened with one or two pins. The
ether is plentifully sprinkled over the interior of the
cone, which is placed over the patient's nose and
mouth, that the vapour may be copiously inspired.

The most perfect method of administering ether is
by Clover's Gas and Ether Apparatus. By this plan,
nitrous oxide is given first to produce partial insensibility. Then gas mixed with ether vapour, and lastly
pure ether. By this method, the patient is put to
sleep more peacefully than by chloroform ; but, as the
apparatus is cumbersome, Clover's Portable Regulating
Ether Inhaler (fig. 128) is more generally applicable
in practice. The object of this instrument is to
induce anesthesia partly by the diminution of oxygen
in the respired air, hence the patient is made to
breath the same air again and again before he inhales

ether. Two ounces of ether, sp. gr. ·720, are put into the reservoir, and the mask fitted to the face. The patient respires deeply into and from the bag for a few seconds, then the ether vapour is gradually allowed to mix with the air inspired, and the bag becomes

Fig. 128.—Clover's Regulating Ether Inhaler.

charged with respired air and ether vapour. In about three minutes anesthesia is produced.

As soon as complete insensibility is induced, a small quantity of fresh air is admitted by lifting the mask from the face while one or two inspirations are taken, and anesthesia is maintained by adjusting the supply of ether or of air, according to the condition of the patient.

Ethidene Dichloride is employed by some operators as an anesthetic ; but the variability of its composition is against its use. It is moreover an objection that the influence of ethidenc on the heart is quite as powerful as that of chloroform.

Bichloride of Methyline has also the same uncertainty of composition ; hence, though still employed, it has never come into general use.

A mixture of four parts of ether with one of chloroform will be found to act as well or better than either of these two compounds. This mixture is especially adapted for operations on the eye. It is administered through Clover's Ether Inhaler in the same manner as simple ether.

Nitrous Oxyde Gas is given alone for operations not requiring prolonged anesthesia, such as drawing teeth, opening abscesses, and the like. Clover's mask, india-rubber bag and tube, and liquid gas bottle are the apparatus employed. Care must be taken that the mask fits closely to the face, for nitrous oxyde gas causes no struggling or excitement when inspired *pure*, but the admixture of a small quantity of common air produces the well-known effects of laughing gas. Insensibility is obtained in about thirty seconds ; by which time the patient, being then not merely narcotised but partially asphyxiated, must be allowed to breathe pure air a few times before the gas is again inhaled.

It is during this short-lived unconsciousness that the operation is performed. If the operation is prolonged for more than a few seconds, the gas may be inhaled again. But the necessity for frequent interruptions to allow the patient to breathe pure air renders nitrous oxyde gas unsuitable for long operations.

Artificial Respiration.—Many plans are employed; but the most efficient are the three to be now described.

Marshall Hall's Method.—Lay the patient on the floor, with the clothing round his neck, chest, and ab-

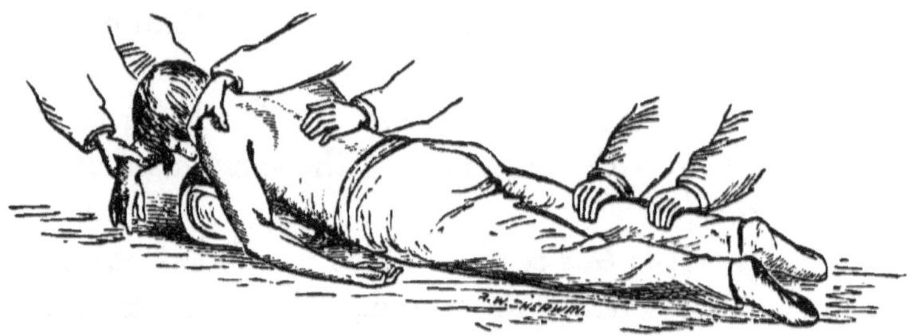

Fig. 129.—Artificial Respiration. Marshall Hall's method. · 1st position .

domen loose; if wet, remove it, and throw over his body a warm blanket. Clear out the mouth, and turn

Fig. 130.—Artificial Respiration. Marshall Hall's method. 2nd position.

the patient *on his face,* one arm being folded under his forehead (see fig. 129), and the chest raised on a folded coat or firm cushion. Next, turn the patient well on to his side, while an assistant supports the head and arm doubled underneath it (see fig. 130), and confines his attention to keeping the head forward and the mouth open during the movements to and fro. When two seconds have elapsed turn the body again face downwards, and allow it to remain so for two seconds, and then raise it as before. This series of movements, occasionally varying the side, should be repeated about fifteen times a minute and *continued until spontaneous respiration is restored, or until two hours have been thus spent in vain.*

Silvester's Method.—Lay the patient on a flat surface, the head and shoulders supported on his coat folded

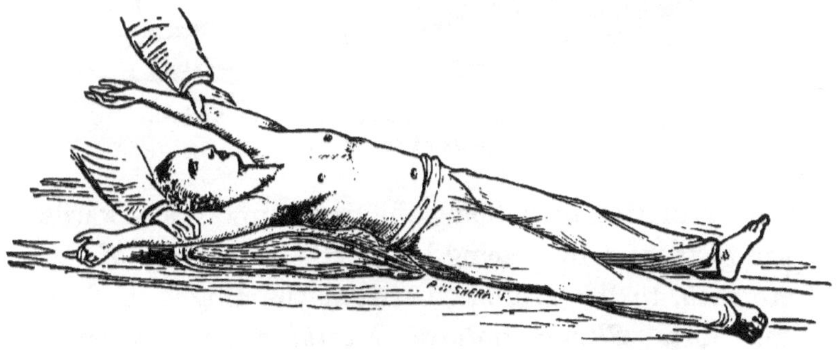

Fig. 131.—Artificial Respiration. Silvester's Method.
Expanding the Chest.

into a firm cushion. Loosen all tight clothing, and if wet replace it by a warm dry blanket, his arms being outside the blanket. Clear the mouth of dirt, blood, &c., draw the tongue forwards, and fasten it to the chin by a piece of string or tape tied round it and the lower jaw. Next, standing at the patient's head, grasp the arms at the elbows, and draw them gently and steadily upwards till the hands meet above the head (see fig. 131); keep them so stretched for two seconds. Then slowly replace the elbows by the sides, and press gently inwards for two seconds (see fig. 132). These movements are repeated without hurry about fifteen times in a minute, until a spontaneous effort to breathe is made, when exertion should be directed to restoring

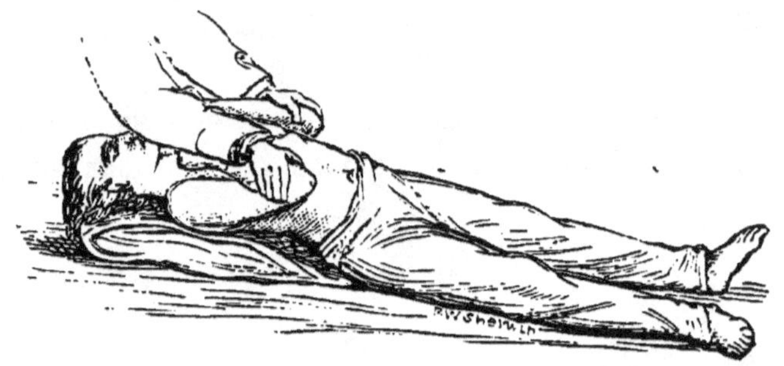

Fig. 132.—Artificial Respiration. Silvester's Method. Compressing the Chest.

the circulation by rubbing the limbs upwards towards the body, and by placing hot bottles at the pit of the stomach, to the armpits, between the thighs, and to the feet. *Should natural breathing not commence, artificial respiration should be continued for two hours before success is despaired of.*

Howard's Method—RULE 1. *For ejection and drainage of fluids, &c., from the stomach and lungs.*

Position of patient.—Face downwards; a hard roll of clothing beneath the pit of the stomach, making that the highest point, the mouth the lowest. The forehead resting upon the forearm or wrist, keeps the mouth from the ground (see fig. 133).

Position and action of operator.—Place the left hand well spread upon the base of the chest to left of the spine, the right hand upon the spine a little below the left. Bear upon the hands so placed, with a forward motion of the body, and with as much weight and force as the age and sex of the patient will justify. End this pressure in two or three

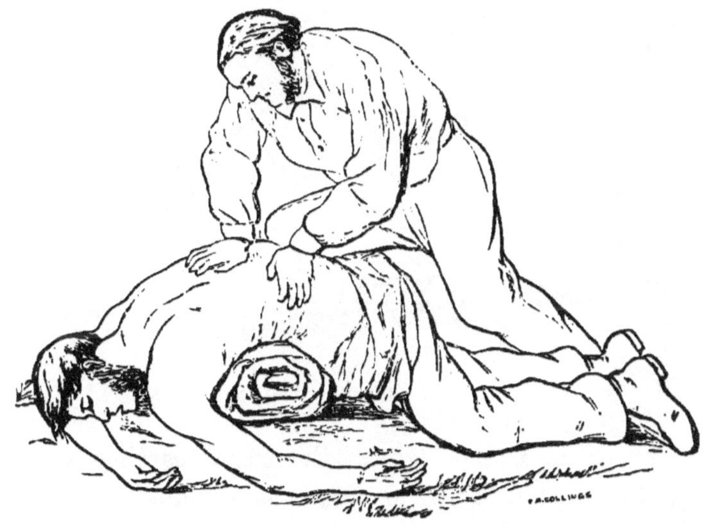

Fig. 133.—Expelling the Water.

seconds with a sharp push which helps to jerk you back to the upright position. Repeat this manœuvre two or three times if matters continue to be ejected from the mouth.

RULE 2. *To perform Artificial Respiration.*

Position of patient.—Face upwards; the hard roll of clothing beneath the chest with shoulders *slightly* inclining over it. Head and neck bent back to the uttermost. Hands on top of head (one twist of a handkerchief round the crossed wrists will keep them there). Rip off clothing from neck and waist (see fig. 134).

Position of operator.—Kneel astride of the patient's hips ; place your hands upon his chest, so that the ball of each thumb and little finger rests on the inner

Fig. 134.—Inflating the Lungs.

border of the free margin of the costal cartilages, the tip of each thumb near or upon the Xyphoid cartilage, the fingers fitting into the corresponding intercostal spaces. Fix your elbows firmly, making them one with your sides and hips ; then—

Action of operator.—Pressing upwards and inwards towards the diaphragm, use your knees as a pivot, and throw your weight slowly forwards until your face almost touches that of your patient. End with a sharp push which helps to jerk you back to your erect kneeling position. Rest three seconds : then repeat this bellows'-blowing movement as before, continuing it at the rate of from seven to ten times a minute. When a natural gasp occurs, take the utmost care to gently aid and deepen it into a longer breath until respiration becomes natural. When practicable, have the tongue held firmly out of one corner of the mouth by the thumb and finger armed with dry rag.

NOTE.—Avoid impatient vertical pushes. The squeezing force must be increased gradually up to the maximum suitable to the age and sex of the patient.

Abandon no case as hopeless without at least one hour's ceaseless effort.

Richardson's Ether Spray Producer (fig. 135)

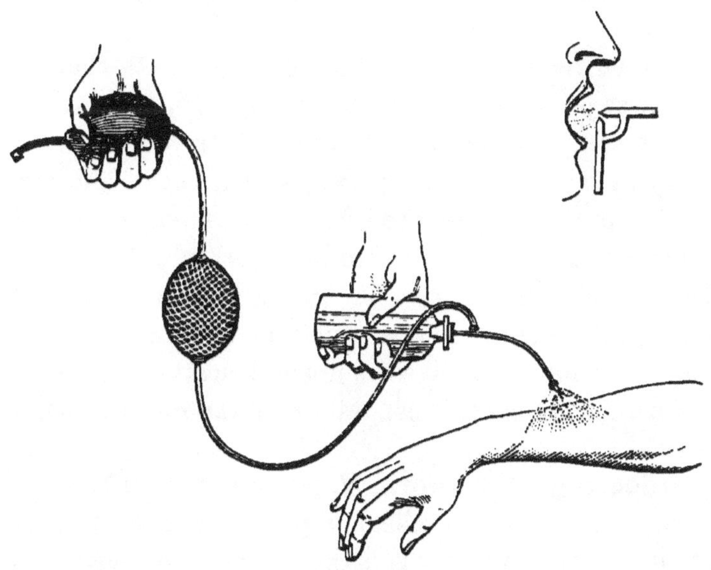

Fig. 135.—The Spray-producer.

consists of a tube on which two india-rubber bags (Higginson's pump) are placed; one, protected by a silk net, acts as a reservoir; the other, furnished with a valve, is the pump; these drive a constant stream of air over the tip of a fine tube projecting from a flask of ether ; this sucks up the ether and throws it in fine spray on the surface to be chilled by its evaporation. The ether for this purpose must be very pure and dry, having a specific gravity of ·720, or the evaporation will not be sufficiently rapid to produce congelation. The first effect of the spray is a numbing aching pain with reddening of the surface. This is succeeded by a pricking pain. In ten seconds, if the ether be good, a dead white hue spreads rapidly over the skin, and when this appears the surface is quite insensible.

The bottle and elastic air-pump may be attached to the glass jet seen in the corner of fig. 135, which then makes an apparatus for injecting astringent solutions in spray over the nasal passages, the throat, and air-tubes ; but the tubes used for watery fluids are much wider than that for pulverising ether into spray. Tannin in solution of 1—5 grains to the ounce of water, sulphate of zinc, or alum in similar quantity, may be thus inhaled with much benefit by persons suffering from chronic congestion of the mucous membranes.

Injecting Chloroform Vapour into the Uterus is a ready means of relieving pain in cancer of that organ ; special apparatus is made for the purpose, but an ordinary elastic clyster syringe will answer the purpose, if the flask is unscrewed and a bit of sponge put into it. A few drops of chloroform are poured from time to time on to the sponge, while air is pumped through the delivery tube, which is passed up the vagina to the ulcerated cervix-uteri.

Subcutaneous Injection.—The syringe for this operation (fig. 136) consists of a graduated glass tube holding from ten to twenty minims. The piston works in a silver continuation of the graduated tube, and is thus kept clear of the solutions used for injection. To the nozzle of the syringe fine sharp-pointed cannulæ are screwed on ; they are of different lengths, some of steel, others of steel gilt ; the gilding renders the points very blunt, and consequently much more painful to insert. In filling the syringe, care should be taken not to draw the fluid above the level of the graduation on the tube, that the exact amount injected may be read off as the liquid sinks in the tube. The finer the cannula, and the sharper its point, the less pain is caused by its introduction. Some persons much dread the puncture ; for them the pain may be entirely prevented by numbing the surface with ether-spray (see page 233) before inserting the syringe, though usually the pain is too trifling for this precaution to be necessary.

The solution of morphia should contain a grain in six or in twelve minims, and be as little acid as possible. In injecting morphia it should be recollected that $\frac{1}{4}$ grain is the usual dose to allay pain, and produce sleep ; doses even far smaller often suffice for this purpose, though very much greater quantities can be administered by injection, where long use has rendered the patient tolerant of the drug.

A spot should be chosen where the skin is loose and has a good layer of fat, the arm for example, and the skin should be steadied by putting it on the stretch with the left thumb and fingers, while the point of the cannula, held at right angles with the surface, is thrust quickly, with a slight rotatory motion, completely through the skin into the subcutaneous cellular tissue. If the fluid is injected into the skin itself, inflammation

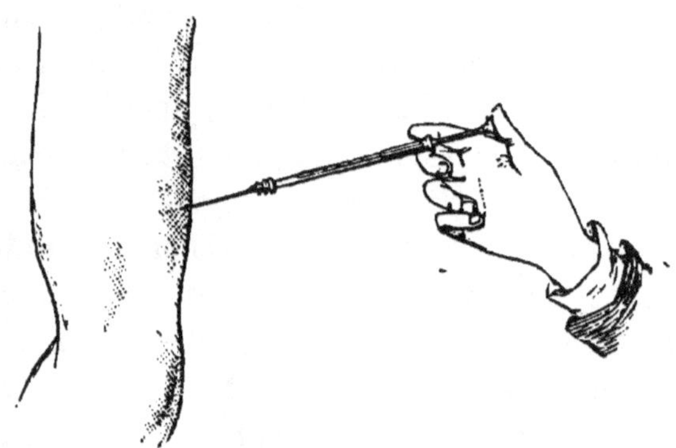

Fig. 136.—Subcutaneous Injection.

and suppuration of the puncture sometimes ensue. After the cannula is withdrawn, the finger should be placed for a few seconds over the puncture, or much of the fluid will leak out again. When large quantities of solution (one or two syringefuls) are injected the cannula need not always be withdrawn, the nozzle can be unscrewed and the syringe charged again ; but more than ten or twelve drops injected into one place generally cause much pain, even where the cellular tissue is very loose.

The syringe and cannula should be carefully cleaned by sending plenty of cold water through them every time they are used, and the point frequently sharpened on a hone, or the cannula will rust and become unfit for use.

Collodion is much used in drawing the edges of small wounds together, &c. Plastic collodion, or the flexible collodion of the British Pharmacopœia, 1867, have advantages over the contractile form by furnishing a tougher pellicle, and yielding to the movements of the skin beneath without cracking. Collodion should be kept for use in a small wide-mouthed bottle, with stopper and brush, and when employed should be laid on quickly in a thick mass, so that the crust it leaves shall be of one layer. A tougher crust is obtained if a piece of muslin is soaked in the collodion and then laid on the wound, than if the collodion is used alone.

Vaccination.—The lymph of the vaccine disease is termed "bovine" lymph when that disease is propagated in the calf, and humanised lymph when propagated in the human subject. In all cases where a child is employed as the source of lymph, it must be carefully examined and its family history ascertained; for no child that is not clearly in good health should be selected for a vaccinifer.

When lymph is needed for use, it is taken between the seventh and tenth days, and preserved on lancet-shaped slips of bone 1 inch long, called *points*. These are dipped in the lymph as it exudes from the vesicle, and exposed to the air till dry; they are then wrapped in paper ready for use. When used, the lymph should be moistened, by holding the point over a vessel of steaming water for a few seconds, before inserting it in the wound made to receive the lymph.

Points often lose the virus in a few days, and should, if possible, be used on the day they are charged. The lymph may be much longer preserved if hermetically sealed in *glass tubes*. These tubes are about the thickness of a darning-needle, 3 inches long, and open at both ends. When the tube is to be charged, one end is inserted in the lymph exuding from a punctured vesicle; a drop then enters the tube by capillary attraction, but filling not more than half its interior; a few shakes of the hand will send the drop a little further in. The lymph end of the tube is then taken in the thumb and forefinger, while the unoccupied part of the tube is passed once or twice quickly through the flame of a candle. This rarifies the air, and while it is warm the end is closed by melting it at the edge of the flame. The second end is then closed in the same way as the first. When the lymph is wanted for use, the ends of the tube are broken, and the lymph blown out on to the point of a lancet. Lymph preserved in these tubes retains its efficacy an indefinite time. The National Vaccine Establishment (care of the Medical Officer of the Local Government Board, London, S.W.), supplies to medical practitioners both points and tubes of humanised lymph gratis on application by post; letters not needing stamps. The bovine lymph may be obtained of Dr. Renner, 228, Marylebone Road, London, N.W., at a moderate charge.

In performing the operation the common lancet does very well; but two or three forms of narrow grooved lancets are employed by surgeons for this purpose. The operation is most successful when the lymph is transferred direct from arm to arm; the lancet making the puncture is then charged at the vesicle of a child vaccinated a week before, and points are unnecessary. When making the puncture the surgeon grasps the child's arm in his left hand and puts the skin on the stretch over the insertion of the deltoid with his left forefinger and thumb. He next pushes the lancet downwards beneath the cuticle, about 1-10th of an inch, to raise a little pocket. He then charges his lancet with lymph and inserts it into the pocket, or if using points, inserts the moistened point for a minute. As he withdraws the point he presses down the pocket on the point with his left thumb nail, that the lymph may be well wiped off the point and left in the wound. This process is repeated at four or five places and the operation is complete. The corium should not be penetrated, or it will bleed freely and the blood will wash away the lymph; one drop of blood is of little consequence; indeed, it shows that an absorbing surface has been reached. On the other hand, in taking lymph from a vaccinifer, the greatest care to avoid drawing blood or anything but the clearest lymph must be exercised.

The phenomena following the insertion of the vaccine virus in an infant's arm are as follows:—On the second day the puncture is slightly elevated; on the third it begins to grow red; on the fifth it is marked by a distinct vesicle with a depressed centre and red areola; on the eighth the vesicle is perfect, of pearl-like aspect, full of clear lymph; the areola, often little marked by the eighth day, rapidly increases on the ninth and tenth days, and reaches an inch or more in diameter. This bright red inflammatory action in the skin is essential to show the system is properly infected with the vaccine disease. By the twelfth day the areola has lessened, the lymph is yellow, and often escapes by rupture of the vesicle ; on the fourteenth day the vesicle has dried to a scab, that falls off on the twenty-first day, leaving a dotted cicatrix, the vestige of the multilocular structure of the vesicle. The three important marks diagnostic of the vaccination being satisfactory, are—1, the pearly multilocular vesicle of the 8th—9th day; 2, the widely-spread areola on the 9th—12th day ; 3, the well-marked foveated cicatrix after the scab has fallen.

Observation shows that the number of people who take small-pox after vaccination is very small indeed, when more than three well-marked scars exist ; and this number at least should be secured by making five insertions of lymph at the time of vaccination.

CHAPTER VI.

SURFACE-GUIDES AND LANDMARKS.

CERTAIN fixed points and marks on the surface of the body have bearings of which a precise knowledge is necessary in investigating injuries and disease.

The Head comprises the cranial and facial regions, the interior of the nose and of the mouth, and the pharynx.

In the *cranial region*, the occipital protuberance, the frontal and parietal eminences, the mastoid process, the zygoma, the margin of the orbit, and the external angular process of the frontal bone, can be always detected. Making use of these :—

1. The occipital protuberance marks the site of the convergence of the venous sinuses in the torcular Herophili.

2. A line drawn from the occipital protuberance through the meatus auditorius externus to the external angular process of the frontal bone and from that point over the orbit to the root of the nose, corresponds to the lower level of the great brain; the posterior lobe being behind the auditory meatus, the middle and anterior lobes opposite and in front of it. The bifurcation of the fissure of Silvius corresponds to

a point on the surface $1\frac{1}{4}$ inches behind, and $\frac{1}{4}$ inch above the level of, the external angular process. Below the posterior part of the line is the little brain and the medulla oblongata and pons. The lateral sinus, starting from the torcular Herophili, passes outwards for more than three-fourths of the distance between the occipital protuberance and the external auditory meatus along this line ; it then turns downwards towards the mastoid process.

3. A line carried directly forwards from the occipital protuberance over the skull to the root of the bridge of the nose, denotes the position of the superior longitudinal sinus.   It also corresponds to the sagittal suture, and crosses the site of the fontanelles.

In a healthy child the posterior fontanelle is closed before birth : the anterior is open during the greater part of the first year, but closes towards the end of the first, or in the course of the second year.

The usual thickness of the skull in health is $\frac{1}{5}$ inch. It is often as thin as cartridge paper at certain points between the eminences, but it may also be $\frac{1}{2}$ inch thick in the healthy state.

4. The exact line of the anterior branch of the *middle meningeal artery* is one drawn through points at equal distances behind the external angular process and above the zygoma : *i.e.*, $1\frac{1}{4}$, $1\frac{1}{2}$, or 2 inches.   The last is the best position for trephining, as below this the vessel is often enclosed in a bony canal.

The positions of the sinuses and of the middle meningeal artery influence the selection of a situation for trephining the skull.

5. The *anterior temporal artery*, the vessel punctured

for bleeding at the temple, can be felt pulsating 1¼ inches behind and above the external angular process of the frontal bone at the edge of the temporal fossa.

6. The *supra-orbital artery* can be felt pulsating at the juncture of the inner and middle thirds of the upper margin of the orbit.

7. The trunk of the *superficial temporal artery* can be felt just in front of the tragus of the ear. Here it crosses the root of the zygoma, and it may be compressed against that bone.

The *occipital artery* can be felt half-way between the occipital protuberance and the mastoid process, at which point it pierces the attachment of the trapezius muscle.

The *external auditory canal* is about 1¼ inches in length in the adult, but much less in infants, owing to the shallowness of the osseous part of the passage at birth : rather less than half being cartilaginous and rather more than half osseous. It is directed inwards and slightly forwards, being also arched with a downward concavity. To straighten the canal for inspection, draw the auricle upwards, backwards, and a little outwards. With a good light, the *membrana tympani* can be seen. When healthy, it is a greyish membrane, slightly receding and placed obliquely forwards and inwards across the canal; so that the outer surface, which is concave, looks downwards and forwards as well as outwards. The handle of the malleus can be descried through the membrane.

In the facial region the bony points which serve as guides are the margins of the orbit, the zygoma, the malar bone, and the margin, angle and ramus of the lower jaw.

1. The supra-orbital notch is to be found at the junction of the inner and middle thirds of the upper margin of the orbit. It marks the position of the *supra-orbital vessels* and *nerve*, and a straight line drawn from it through the interval between the lower bicuspid teeth crosses the *points of issue through bony foramina of the three principal facial branches of the fifth cranial nerve.* Sometimes these are divided to relieve pain in tic douloureux. Below the base of the zygoma the joint of the lower jaw can be easily felt.

2. The *external carotid* artery mounts to the side of the head just behind the ramus of the lower jaw. It must be avoided in incisions or in puncturing abscesses.

3. The *facial* artery crosses the margin of the lower jaw just in front of the masseter muscle, and here it may be compressed. The artery can be felt to pulsate also near the angle of the mouth and the ala of the nose.

4. At the inner corner of the orbit is the guide to the lachrymal sac : namely, the lower border of the *tendo oculi,* which is visible when it is made tense by drawing the lids outwards when the sac is to be opened. The *punctum lachrymale* is also easily seen at the inner end of each eyelid, at the apex of the minute *papilla lachrymalis;* when passing a probe along the duct, the lid should be drawn slightly outwards : and it should be remembered that the course of the duct is at first vertical and afterwards horizontal.

5. A line drawn from the bottom of the lobe of the ear to a point midway between the ala of the nose and the corner of the mouth denotes the course of the parotid duct, and some large branches of the facial nerve. The duct (Stenson's duct) is often, however, found at a higher level ; it opens into the mouth, opposite the second upper molar tooth.

The greater part of the *parotid gland* lies immediately below the ear, but besides deeper prolongations a superficial part reaches to the front of the masseter muscle. In incising abscesses in this part, the knife should be carried horizontally forward to avoid cutting the twigs of the *pes anserinus* of the facial nerve.

The course of the trunk of the facial nerve is indicated on the surface by a line drawn downwards and forwards from the anterior border of the mastoid process at the point where that border meets the ear.

The Cavity of the Nose.—When inspecting this cavity, push the head backwards and raise the tip of the nose. The parts to be seen are :—

1. The septum narium. This is said to be occasionally perforated by a small hole in health. When perforated by disease, the hole, if small, is usually at the junction of the cartilage and the bone ; and it enlarges chiefly at the expense of the bone.

2. The inferior spongy bone.

3. The lower and middle meatuses. In the lower meatus $\frac{1}{4}$ inch behind the bony margin of the nostril, overhung by the inferior turbinated bone, the *ductus ad nasum* opens. It can be reached and entered by a

probe of which the last half-inch is bent to a semicircle and the remainder to a larger curve in the opposite direction like an italic *f.* The opening is very small, and guarded by a fold of mucous membrane. Polypi can be seen as bluish-pink bodies blocking the passage of the inferior meatus. When growing entirely in the nose, they start from the superior turbinated bone, but they may invade the cavity of the nose from the antrum of Highmore, or from the pharynx.

Irregular bends in the cartilage of the septum, caused by injuries to the nose, sometimes make projections in the meatus, which are mistaken for polypi. The real nature of the projection is easily detected by observing that the obstruction in one meatus is compensated by the widening of the other meatus.

In the Mouth—when wide open and the tongue laid back—the hard and soft palate are to be seen, at the junction of which in the mesial line a natural depression or seam sometimes exists.

The *tonsils* are placed between the pillars of the fauces. In health they do not project beyond the arches. They are separated by a thin fascia from the internal carotid artery, so that, when cutting or lancing them, the point of the knife should be directed obliquely towards the centre of the pharynx, while the swollen gland itself is drawn forwards and inwards. Externally the angle of the lower jaw corresponds to the position of the tonsil.

If the angle of the mouth be drawn aside the interior of the cheek can be examined, and the papilla in which is placed the orifice of Stenson's duct may be seen opposite the second molar tooth of the upper jaw.

If the tongue is raised, on its under surface the *ranine veins* are seen on each side of the mesial furrow. They indicate the position of the ranine arteries. The frenum linguæ passes from the tongue to the jaw in the mesial line. When it is prolonged to the tip of the tongue, and is to be snipped to set free that member, a blunt-pointed scissors should be used, and the points directed towards the jaw, to avoid the blood vessels of the tongue. A ridge passes outwards from the frenum to the floor of the mouth; here ducts of salivary glands open. The papilla on which Wharton's duct opens is easily visible very near the middle line; a fine probe may be passed into it for an inch or more. Beneath it the sublingual gland is placed and ranulæ form.

The finger in the mouth can detect the following :—

1. The *tuberosity of the superior maxilla.*

2. The *coronoid process* of the inferior maxilla. Between these points is a deep depression, in which a deep temporal abscess may point and be incised.

3. The hamular process and, descending from it, the *pterygo-maxillary ligament.*

4. The *gustatory nerve* lies just below the last molar tooth of the lower jaw, near the pterygo-maxillary ligament. As the finger is passed over the nerve, a thrill of heat and pain is felt in the mouth. At this point the nerve is sometimes cut across to relieve the pain of cancer in the tongue.

However tightly the jaws are set in tetanus, a

flexible catheter can always be passed behind the teeth into the mouth.

The finger may be passed over the tonsils to the pharynx, the greater part of the walls of which can be reached. If the finger be turned upwards round the soft palate, the posterior nares, and the orifices of the Eustachian tubes may be examined. The parts of the vertebral column to be felt are, the basilar process of the occipital bone at the top of the pharynx, the anterior margin of the body of the atlas opposite the lower margin of the posterior nares, and the axis opposite the soft palate.

Certain tumours, such as nasal or pharyngeal polypi, gummata of the palate and pharynx, abscess in the tonsils, retro-pharyngeal abscess and impacted foreign bodies, may also be detected. The largest artery in the palate (posterior palatine) runs forward from a point half-an-inch internal to the wisdom tooth.

The Neck may be divided into *anterior*, *lateral*, and *posterior* regions. The anterior reaches on each side from the middle line to the sterno-mastoid muscle ; the lateral from the sterno-mastoid to the trapezius muscle, in a backward direction, and downwards to the collar-bone. The posterior region extends from the occiput to the seventh cervical spine.

In most male adults the various landmarks are conspicuous, but in women and children the neck is smooth and rounded ; therefore to bring the several marks into relief, where the anterior region is examined, the head should be thrown back over a small pillow placed behind the shoulders.

In this position the wind-pipe is drawn half an inch higher above the sternum, and the carotid arteries are brought nearer to the surface.

Keeping to the middle line, the most conspicuous landmark is the *pomum Adami*, angular and prominent in most men, but rounded and only slightly projecting in women and children. The rounded upper border, with the central notch, is readily detected.

The *superior thyroid artery* lies on the upper part of the lobe of the thyroid body, and can be felt pulsating outside the thyroid cartilage. It is often wounded in cases of " cut throat."

A little above this edge the *body of the hyoid bone* can be felt, its great horns extending on each side. The *lingual artery* runs just above the top of the great cornu of the hyoid bone. It is often the source of dangerous hæmorrhage in cases of cut throat. In a muscular neck, with small amount of fat, the anterior belly of the digastric muscle can be defined, passing from the body of the hyoid bone to the chin. In the hollow between the horn of the hyoid and the border of the jaw, the sub-maxillary salivary gland can be felt or even seen in thin persons. The lymphatic glands of this region are affected and enlarged by irritation of the lips and chin, and of the floor of the mouth.

Between the hyoid bone and thyroid cartilage the finger feels *the thyro-hyoid membrane*. This is frequently severed in cases of cut throat; the greater part of the epiglottis lies above the incision when this is carried through that membrane ; but the narrow part of it is divided, and the vocal cords are visible below the cut.

Passing downwards in the middle line from the pomum Adami, first a depression, and then a hard smooth surface, are felt; the hard surface is the *cricoid cartilage*, the depression is the *crico-thyroid* membrane. Through this membrane the wind-pipe is opened in laryngotomy.

The *cricoid cartilage* can be felt in both sexes at all ages. It is placed opposite the fifth cervical vertebra. The cricoid cartilage is a landmark in tying the common carotid artery, and in examining the gullet, which commences behind the cricoid; and here foreign bodies, too large to pass to the stomach, lodge, and may be felt. In œsophagotomy, the cricoid cartilage is a necessary point of reference.

In *tracheotomy* the cricoid cartilage is the guide for the trachea (which is small and very movable in children), for the second, third, and fourth rings are covered by the *isthmus thyroideœ*, and the lower rings are too deeply placed to be felt by the finger. The isthmus often lies over the first ring and even over part of the cricoid, especially in children. At the level of the sternum the trachea is one inch below the surface, and in adults considerably more, but its depth varies with the amount of fat present in the individual.

The *isthmus thyroideœ* lies immediately below the cricoid. It is indistinct in men, but in women is often easily made out as a soft narrow band, even when the thyroid body is not enlarged. On each side of the trachea and of the cricoid and thyroid cartilages, *the lobes of the thyroid body* are placed. Being attached to the larynx, this body rises with the larynx when the act of swallowing begins, and falls to its usual position at the end of that act. Hence an enlarged thyroid body (bronchocele) is distinguished from other tumours of the neck by two characteristics. (*a*). The mass rises and falls during deglutition. (*b*). The superior thyroid artery, coursing forwards and downwards at the level of the thyroid cartilage, is usually larger and always more prominent when the underlying thyroid body is enlarged. Tumours of the neck not originating in the thyroid body may be pushed aside, but do not rise, during deglutition, and if they affect the superior thyroid artery, they conceal it.

The *sterno-mastoid* muscle is in all persons easily defined, and is the landmark for several important structures. It separates the anterior from the lateral region of the neck. The two sternal attachments, converging, form the deeply marked *supra-sternal notch*.

Near the anterior border the *anterior jugular vein* is usually to be found at the lower part of the neck; above, it begins at the middle line near the chin. At the lower part of the supra-sternal notch the veins of each side are commonly united by a communicating branch. Each vein then winds backwards closely behind the attachment of the sternal head of the sterno-mastoid. It may be wounded in division of the sterno-mastoid muscle for the cure of " wry neck."

The anterior border of the muscle overlies the *common*, the *external* and the *internal carotid arteries*. The position of this vessel, as far as its bifurcation opposite the upper border of the thyroid cartilage, is indicated by a line drawn from the sterno-clavicular joint to a point midway between the mastoid process and the angle of the lower jaw, being least overlaid by muscle above the cricoid cartilage.

The pulsations of this vessel can be felt along the whole of its length above the sternum. It may be compressed by the thumb against the transverse process and body of the sixth cervical vertebra, though in doing this it is difficult to avoid pressing on the internal jugular vein and pneumogastric nerve. The landmarks for the preliminary incision, when about to tie the artery in the usual place, are : the anterior border of the sterno-mastoid, the angle of the jaw, and the cricoid cartilage.

The *internal jugular vein* lies behind the interval between the sternal and clavicular parts of the muscle.

The *lateral region* of the neck is bounded in front by the clcido-mastoid muscle, behind by the trapezius, and below by the clavicle. It is marked on the surface by a depression (*supra-clavicular fossa*) that varies in size according to the width of the clavicular attachment of the defining muscles. This region can be most readily explored when the elbow is supported on a chair-back or table, and the neck slightly bent to that side. This position, by relaxing the muscles and fascia, permits the finger to explore the deeper parts of the region.

The *subclavian artery* is felt a little above the clavicle, just outside the border of the clcido-mastoid muscle, where it forms an arch, into the concavity of which rises the pleura and apex of the lung. The vessel is here passing over the first rib, against which it may be compressed. To do this the thumb or the ring of a door key, round which a strip of lint is wound as padding, is thrust downwards and backwards just outside the origin of the cleido-mastoid muscle. This point of the cleido-mastoid denotes the *tubercle of the first rib* and the *anterior scalenus muscle*, land-marks required in cutting down upon the subclavian artery.

The *external jugular vein* crosses the sterno-mastoid usually about the middle of that muscle, and enters the lateral region to reach the subclavian vein. Its course is tolerably well defined by a line drawn on the skin from the angle of the jaw to the middle of the clavicle. If the finger be pushed into the depression, just above the middle of the clavicle, the vein is usually compressed, and by filling out becomes con-spicuous. When this vein is cut, to let blood, the in-cision is made directly upwards at the part which over-lies the sterno-cleido-mastoid muscle. This direction is somewhat oblique to the course of the vein, but it is also across the fibres of the platysma muscle, which must be so cut that they may gape, and not obstruct the flow of blood.

In a long thin neck the *omo-hyoid muscle* is visible during forced inspiration, nearly at the level of the cricoid cartilage, making tense the cervical fascia. In such a neck the cords of the brachial plexus may also be seen and felt.

The lymphatic glands in the lateral region are extremely numerous and are frequently enlarged. Several form a chain contained in a fold of the fascia, just behind and beneath the sterno-mastoid, where they are easily detected.

Besides enlarged lymphatic glands, other swellings are met with : among which are cysts of the cellular tissue of the neck, abscesses around the lymphatic glands, solid tumours pressing on the carotid or subclavian arteries, or aneurism of the carotid or subclavian arteries, and gummata in the substance of the sterno-mastoid (the latter are usually near the sternal end of the muscle). Abscess from the cellular tissue between the deep vessels of the neck and the pharyngeal muscles sometimes wanders forwards until it points in front of the sterno-mastoid in the upper part of the neck. In defining the origin and nature of any swelling in this region, its relations to the landmarks just enumerated must be ascertained.

The posterior region, or nape of the *neck*, extends from the occiput and superior curved line to the spine of the seventh cervical vertebra. Along the mesial line is a vertical depression, of which the borders of the trapezius form its lateral boundaries. In this, below the occiput, the spine of the axis can be felt, and also at the lower end of the depression the last three cervical spines through the skin. The most prominent spine is the first dorsal. At the upper end the nuchal lymphatic glands enlarge with irritation of the scalp. In weakly persons the cellular tissue suppurates about the glands, when the surface is irritated by pediculi or impetigo, and abscess forms at the nucha.

The interval between the sixth and seventh cervical spines corresponds to the termination of the pharynx in the gullet and of the larynx in the trachea.

The thick skin of this part is a favourite seat of boils and carbuncles, and, at the lowest part, of fatty tumours. The large number of short fibrous connections between the skin and deep fascia here, render the skin less movable, and greatly interfere with the excision of tumours.

The Thorax has some landmarks to be detected in all persons, and others which cannot be made out in the very fat.

Those to be *always* made out in front are, the clavicles, the sterno-clavicular joint, the sternal notch, the joint between the first and second pieces of the breast-bone, and the nipples.

Those which can be *generally* felt are, the sternum, the ensiform cartilage, the costal cartilages except the first, the lower ribs, the apex beat of the heart, and the interval between the great pectoral and the deltoid muscle which forms a depression just below the clavicle in which the coracoid process is felt.

The nipple in the *male* lies usually over, or just below the fourth rib. In the female it varies much.

In the axillary and infra-axillary regions of the chest, the following landmarks are *constantly* to be made out :—

The borders of the pectorals, and of the latissimus and teres major muscles forming the anterior and posterior boundaries of the arm-pit, and the two last ribs.

In addition to these some more can *generally* be detected.

If the arm is raised from the body, 4 digitations of the serratus become prominent. They correspond to the 6th, 7th, 8th, and 9th ribs.

In thin persons the ribs and intercostal spaces can be felt and seen between the pectoral and latissimus, as well as the cartilages of the lower ribs which form the border of the chest-wall.

The most prominent part of the *forward curve* of the clavicle marks the last part of the *subclavian artery.*

The *inner third* of that bone overlies the *subclavian vein.* It has been lacerated by fracture of the clavicle from direct violence.

The *inner end* of the clavicle marks the *confluence of* the *subclavian* and *internal jugular veins* into the innominate vein.

The *sternal notch* marks the upper border of the first piece or manubrium sterni, as the cartilage of the second rib does the lower border.

Behind the sternal notch in the middle line the *trachea* enters the chest 1 inch or more below the surface ; and, passing deeply behind the great vessels, it divides into *bronchi* beneath the aortic arch opposite the joint between the handle and the blade of the breast bone.

Behind the *upper part of the manubrium* are the commencements of the *innominate* and *left carotid arteries* from the highest part of the aortic arch ; and the *left innominate vein.*

Behind to the *right side of the manubrium,* but

separated by a small interval, are the *right innominate vein* and the upper part of the *superior vena cava*.

The *sterno-clavicular joint* of the *right* side at its upper margin marks the bifurcation of the innominate artery into *right subclavian* and *common carotid arteries*: that of the *left* side marks the *left common carotid* artery.

The *junction of the first and second pieces of the sternum* marks the second *costal cartilage*, an important landmark in ascertaining the area of the heart. It is placed opposite the 5th dorsal vertebra.

The *heart's area* in the front of the chest: *two-thirds* are to the left of the middle line, *one-third* to the right.

Roughly defined, the heart lies behind the 4th, 5th, and 6th left cartilages, and that much of the sternum and intercostal spaces which is on a level with those cartilages. But the position of the organ is altered by the movements of the body, the amount of air in the lungs, and of blood in the heart. Hence it varies somewhat from the following exact dimensions, which are taken from an adult in an erect position, breathing quietly.

The *upper edge of the auricles* is marked by a horizontal line at the level of the 2nd costal cartilages. The *right auricle* extends for *one inch to the right* of the sternum.

The *auriculo-ventricular sulcus* is denoted by a line from the sternal end of the *sixth right cartilage*, to the *third left cartilage*, half way along the cartilage.

The lower border of the *right ventricle* is marked by a nearly horizontal line from the *sternal end of the seventh right cartilage* to the apex.

The *upper* border of the *left ventricle* is marked by an oblique line drawn from the *second left costal cartilage* to a point 1½ *inches below the left nipple* (*i.e.*, in the fifth interspace).

The *apex* is situated 3½ inches to the left of the middle line in the fifth interspace, close to the upper margin of the sixth rib, *i.e.*, 1½ inches below the nipple, and about ¾ inch to its sternal side.

The (*tricuspid*) opening from the right auricle to the ventricle is behind the lower part of the sternum, opposite the fourth cartilage.

The (*mitral*) opening from the left auricle to the ventricle is behind the left side of the sternum, opposite the fourth interspace.

The *aortic orifice* is placed close to *the third left cartilage and behind the sternum.*

The *pulmonic orifice* is a little higher and more superficial than the aortic. It is placed behind *the left border of the sternum and the third cartilage.*

If the area of the heart's dulness does not rise above the second cartilage, there is no pericardial effusion of any consequence.

The heart would be involved in a wound traversing the chest above the *sixth rib.* The *lungs* would be injured by a wound traversing the chest above a line carried obliquely backwards and downwards from the sixth cartilage to the tenth or eleventh rib.

The *line of the arch of the aorta* can be defined on the surface in only the anterior part of its course.

It begins in the *third left interspace* close to the sternum, and ascends behind that bone to the upper border of the *second right cartilage*. Here it curves to the left till it reaches the *middle of the manubrium sterni*, its highest point, thence it crosses backwards to the *left side of the body of the fourth dorsal vertebra*, and descends to the lower border of the body of the *fifth* dorsal vertebra, where it takes the name of the descending aorta.

The *area of the lungs*. The *apices* reach, as a rule, into the root of the neck for $\frac{3}{4}$ inch above the inner end of the clavicle, and $1\frac{1}{2}$ inch above the first rib. In emphysematous persons they can be seen during forced expiration bulging even higher upwards.

The *anterior* borders of the lungs : That of the *right lung* starting from the *root of the neck* at $\frac{3}{4}$ inch above the *inner end of the clavicle*, reaches the *middle line* behind the *lower half of the manubrium*, and con tinues along the middle of the sternum to the *sixth cartilage*.

The anterior border of the *left* lung commences at the apex above the inner end of the clavicle, but does not quite reach to the mesial line behind the sternum. Opposite the fourth left cartilage the border leaves the sternum, and passes outwards to the junction of the fifth rib and cartilage, where it turns sharply back across the fifth interspace, to the middle of the sixth cartilage, where it joins the inferior border or base. The ➤ -shaped notch thus formed allows the apex of the heart to reach the thoracic wall uncovered by lung.

The surface of the diaphragm is arched on its right half, thus the base of the lung descends in front of the liver in the fifth interspace.

The *base* of the *right* lung corresponds to a line drawn on the surface from the side of the sternum opposite the sixth cartilage, along the sixth cartilage, then downwards and backwards across the sixth, seventh, and eighth ribs and interspaces to the ninth rib. At the ninth rib it corresponds to the posterior fold of the axilla, and passing thence to the back, it descends to the level of the tenth dorsal spine.

The *base* of the *left* lung, beginning about the *middle* of the *sixth left cartilage*, passes outwards along the sixth cartilage, and trending downwards to the *tenth* rib in the infra-axillary region, falls to the *eleventh* rib in the back.

The *posterior borders* of the lungs correspond pretty nearly to the vertebral grooves in the back from the first dorsal spine to the twelfth dorsal spine. The pleura reaches lower ; its limit usually corresponds to the twelfth rib and eleventh dorsal spine.

The *attachment of the diaphragm* to the thorax corresponds to a line on the surface drawn from the ensiform cartilage along the cartilages of the seventh and succeeding lower ribs to the first lumbar spine. The arching of the diaphragm in the normal condition corresponds to a double arched line starting on each side from the ninth costal cartilage and rising to the fourth interspace on the right side, and to the fifth rib on the left ; expiration or enlargement and distension of the abdominal viscera may raise this arch. The liver dulness on the right, and the heart's apex-beat and the stomach-note on the left, are additional signs of the position of the diaphragm. *Forced expiration* brings the arch of the diaphragm to the third interspace

Forced inspiration carries it down to the level of the ensiform cartilage in front and of the tenth rib behind. In disease, Walshe records that it may be carried as high as the second interspace, or depressed below the floating ribs.

The *internal mammary artery* reaches the intercostal spaces below the first rib, and courses downwards behind the cartilages near the sternum to the sixth interspace, where it divides into two large branches; the inner branch anastomoses with the epigastric artery. The vessel might be wounded by a stab or torn in fracture of the sternum, and bleed freely; its cut ends may be tied in the three upper interspaces.

In the *lateral* or axillary and infra-axillary regions, several operations may be performed for which acquaintance with the landmarks is necessary.

Abscesses in the axilla collect beneath the stout fascia which forms a floor for that region. Incisions through it should be made vertically, and *midway between* the *pectoral and latissimus dorsi muscles*, where the fascia is separated for a considerable extent from the wall of the chest, to avoid wounding the large vessels which are contiguous to these borders (the external mammary and the axillary vessels to the arm).

A circle carried horizontally through the nipples crosses the sixth interspace in the infra-axillary region.

The lower border of the pectoralis corresponds to the fifth rib. The depression below the clavicle between the deltoid and pectoral muscles marks the axillary artery in its first part and the coracoid process. It is filled by the head of the humerus in subclavicular dislocation.

The *axillary* artery can be felt in the arm-pit for nearly the whole of its course from the lower border of the first rib to the lower border of the teres major muscle. It may be compressed below the clavicle against the second rib by pressing the finger into the infra-clavicular fossa. Also, when it has left the arm-pit, by pushing it against the neck of the humerus.

In *tapping* the chest the infra-axillary region is generally preferred, as it is not thickly covered by muscles. The sixth or seventh interspaces are commonly selected, but the trocar may be inserted at any point of the chest wall, if fluid is ascertained to be behind that point.

Hepatic abscesses may also be opened in the infra-axillary regions.

In the dorsal region of the thorax there are some landmarks of importance readily made out. The spines of the vertebræ can always be felt, owing to the close attachment of the skin to them. They lie in the *spinal furrow;* a groove that is deepest in the lower dorsal and lumbar regions, and may be levelled up in the dorsal and sacral regions in persons of scanty muscle. The position of the spines may be made evident to the eye by rubbing the skin over them a few times with the finger, when a pink spot appears over each spinous process for a few minutes; or by touching the skin with a piece of charcoal.

The *seventh cervical* spine is a prominent landmark from which the rest may be counted; it is, nevertheless, not quite so prominent as the first dorsal spine. Some are denoted by other marks. Thus the scapula overlaps the ribs between the second and the ninth spines. The spine of the scapula, always to be felt, corresponds to the third dorsal spine. When the arm is raised, the inferior angle of the scapula is level with the seventh rib.

The *third dorsal* spine corresponds to the termination of the aortic arch and the *fourth* to the bronchi.

The *angles of the ribs* can be made out in even moderately fat persons. Between them and the vertebral spines lie the vertebral grooves which are occupied by muscles.

Abscesses formed around carious transverse processes or laminæ of the dorsal vertebræ, often point at the outer borders of the trapezius and latissimus dorsi, the strong fascial envelope of those muscles attached to the spinous processes and tubercles of the ribs, diverts the pus from pointing in the vertebral groove.

These muscles can be readily defined in most persons. The anterior border of the trapezius limits posteriorly the lateral region of the neck; the lower border extends from the lower dorsal spines to the spine of the scapula. The latissimus dorsi forms the posterior border of the axilla. Both arise from the spinous processes, and are directed outwards and upwards. Opposite the base of the scapula the muscles are separated, leaving subcutaneous a small part of the rhomboideus major, and of an intercostal space. Through this area, which is comparatively free from muscles, the pleura may be opened when it is desired to drain the cavity at the hinder part.

Origin of Spinal Nerves.

The *eight* cervical nerves rise from the portion of the cord which extends from the *occiput* to *the sixth* cervical spine.

The *six upper dorsal* nerves arise opposite the seventh cervical and four upper dorsal spines.

The *six lower dorsal* nerves arise from the cord opposite the fifth and following spines down to the *tenth dorsal* spine.

The *five lumbar* nerves arise opposite the *eleventh and twelfth* dorsal spines.

The *five sacral* nerves arise opposite the *first lumbar* spine.

The *spinal cord* ends at the upper border of the *second lumbar* vertebra in the *cauda equina*.

The trunks forming the *cervical* plexus, viz., the anterior branches of the three upper nerves, and a great part of the fourth, lie opposite the four upper cervical vertebræ.

Those of the *brachial* plexus, viz., the anterior branches of the fifth, sixth, seventh, and eighth cervical, with the first dorsal and part of the fourth cervical nerves, are placed in the lower part of the neck, between the fifth cervical vertebra and the coracoid process.

The exact seat of an injury to the spinal cord due to fracture of the vertebræ, or other cause, can often be diagnosed by ascertaining the particular part of the skin which is insensible, or the groups of muscles which are paralysed.

The two *internal cutaneous* nerves of the brachial plexus are formed from the first dorsal and eighth cervical trunks.   They supply no muscles, but are distributed to the skin of the inner side of the arm and of the forearm.

The *median and ulnar* nerves are derived from the first dorsal and eighth and seventh cervical nerves. The median has also fibres from the sixth and fifth cervical. Together they supply sensibility to the palm, the *front* of the wrist and fingers, and the *back* of the little and ring fingers. They also supply all the flexors of the wrist and fingers and the pronators of the forearm.

The *musculo-spiral* nerve is formed of fibres from the eighth, seventh, sixth, and fifth cervical nerves. It supplies sensibility to the *outer* and *posterior* aspects of the hand, forearm, and lower part of the arm. It also supplies all the *extensors* of the hand, wrist, and elbow, and supinators of the forearm.

The *musculo-cutaneous* nerve has a higher origin than the two preceding. It is composed of fibres from the seventh, sixth, and fifth cervical nerves. This nerve supplies the skin of the back of the lower part of the forearm and the ball of the thumb. The muscles it supplies are the flexors of the elbow, viz., the biceps and brachialis anticus, and the coraco-brachialis.

The *circumflex* nerve arising from all four lower cervical trunks, supplies the skin over the lower part of the shoulder, and upper part of the arm on the outer side. It is distributed to the deltoid and teres minor muscles.

The *subscapular* nerve arises from the same trunks, and supplies the subscapular, teres major, and part of the latissimus dorsi muscles.

The two *anterior thoracic* nerves arise from all five

trunks, that for the great pectoral muscle coming from the seventh, sixth, and fifth trunks ; that for the small pectoral from the first dorsal and eighth cervical trunks.

The *posterior thoracic* nerve arising from the sixth and fifth trunks supplies the serratus magnus, an important accessory muscle in respiration.

Thus injuries of the cord *below* the *first dorsal* vertebra do not paralyse the upper limbs, though owing to many of the intercostal and some of the accessory respiratory muscles being paralysed, breathing is impeded, and becomes mainly *diaphragmatic*. If the spinal marrow be injured *below* the *sixth cervical* vertebra, the parts supplied by the *cervical* plexus would not be affected, nor the greater part of those supplied by the *brachial* plexus, most of the trunks of which leave the spinal cord above that vertebra. Injury at the level of the *third cervical* vertebra would destroy the phrenic nerve which rises mainly from the *fourth* cervical trunk, and, by stopping all respiratory movement at once, would cause immediate death.

The Abdomen.

Before an examination is made of the abdomen, the patient should lie horizontal, with the head and shoulders slightly raised, and with a pillow or bolster placed under his knees, to relax all fasciæ. The patient should be encouraged to talk, or bidden to hold his mouth open, that the diaphragm and muscles of the abdominal wall may not be fixed. If spasm is present, chloroform should be given. In feeling for the viscera, apply the flat hand, *not* the tips of the fingers, which excite muscular contraction, and press gently inwards; the solid viscus or the posterior wall of the belly will thus soon be reached and recognised.

These *bony* landmarks can always be made out. The *last two* ribs and the cartilaginous margin of the thorax. The *iliac crest* terminating in front at the anterior superior spine. In the middle line below, the *symphysis pubis* can be felt by tucking the finger under the subcutaneous fat, and pushing upwards the thin skin of the root of the penis. When the symphysis is reached, the *spine of the pubes* can be detected by carrying the finger a little outwards. The *ensiform cartilage* can be made out in all but the very fat.

There are also *surface marks* of the abdominal wall. The belly is generally more or less convex in front; at the sides between the iliac crest and ribs it is usually depressed. Along the middle line is a slight groove, which marks the *linea alba* and the interval between the two recti abdominis muscles below. In it the *umbilicus* is a landmark always present. Towards the thorax the groove widens into a hollow, the epigastric fossa, or "pit of the stomach." At the lower end, the groove ceases a little above the pubes in the *mons veneris.* On each side of the groove is the projection of the rectus muscle; in well formed bodies there are two or three transverse depressions, corresponding to the *lineæ transversæ.* One is *opposite* the umbilicus, one *near the ribs,* and one between them *three or four inches above the navel.* Occasionally there is a fourth linea *below* the navel. Between the lineæ transversæ the contracted muscle may form swellings, that are sometimes mistaken for abscess. *Abscess* does also form in the sheath of the rectus, beneath the muscle, and if the obscure swelling it produces is hastily examined, it may be mistaken for tumour within the belly. In thin bodies where spasm is considerable, the recti become clearly defined, and the muscular fibres of the obliqui rise into distinct and tolerably firm cords under the fingers.

Between the anterior iliac spine and the pubic symphysis, *Poupart's ligament* can be traced, generally with a few inguinal lymphatic glands lying along it.

At the pubic spine the *external abdominal ring*, with the cord issuing from it in the male, can be easily felt. In health it usually admits the tip of the little finger.

The linea alba. This underlies the groove on the surface, and has much surgical importance, owing to its being traversed by no large blood-vessels or nerves. It occupies the middle line from the ensiform cartilage to the pubes. It is thin and usually narrow, though in weakly persons it sometimes bulges forwards widely between the recti. The gap thus formed is then plainly visible. The *umbilicus* is placed in the linea alba, a little below the halfway point of the line, and about the level of the *third lumbar* vertebra. The peritoneum closely invests its posterior aspect throughout, except when the distended bladder rises above the pubes.

The *following viscera* are *crossed* by the linea alba proceeding from above downwards. The *left lobe* of the liver, the *stomach, the pancreas and solar plexus,* the *transverse* colon : these are above the umbilicus.

Below the umbilicus are *the small intestines, with the mesentery* covered by the *great omentum.*

A dull percussion note along the linea alba may be produced by the ascent of other viscera above their usual position.

The *urinary bladder*, when completely distended, rises behind the pubes, and separates the peritoneum from the abdominal wall for *about two inches.* The bladder may rise much higher, even midway between the umbilicus and the ensiform cartilage when unnaturally distended by retention ; but the peritoneum then intervenes in a parietal and visceral layer between the wall of the belly and the viscus.

The *gravid uterus* also rises along the linea alba during gestation, and the degree of elevation marks the age of the pregnancy. At the *third* month the fundus is level with the top of the pelvis. By the *end* of the *fourth* month it is *two fingers' breadth above the pubes.* In the *fifth* month the fundus is *half way to the navel.* By the *sixth* month the fundus *is level with the navel.* In the *seventh* month it is *two inches above the navel.* In the *eighth* month the fundus is *two fingers' breadth below the epigastric fossa.* At the end of the *ninth* month the uterus has gained its highest elevation ; it *fills out the pit of the stomach*, and presses the lower ribs outwards. During the *tenth* month the fundus falls slowly, till at the end of gestation it is *two inches below the ensiform* cartilage.

Besides these healthy enlargements of the viscera ; in disease, ovarian, uterine, and, for a short distance, even prostatic tumours, may rise out of the pelvis along the linea alba. Other tumours, such as aneurism, cancer, fœcal accumulations, though the latter are more common in the flanks, may be felt in the abdomen behind the linea alba.

Umbilical hernia protrudes at the umbilicus in the linea alba; hernia at any other part of the interval between the recti being *ventral*.

The following operations are performed by incisions in this line :—

1. The most common, tapping the peritoneal cavity for relief of ascites.

2. Gastrotomy, for relief of internal strangulation.

3. Herniotomy for umbilical hernia.

4. Ovariotomy.

5. Cæsarian section.

6. The high operation for stone.

7. Tapping the bladder above the pubes, for retention of urine.

In the two last operations the peritoneum is not opened.

Regions of the abdomen.   The belly is divided arbitrarily into a fixed number of *regions*, in which the viscera are described as being situated.   The limits of the regions are imaginary, and are represented on the surface by lines drawn through certain fixed points.

Two *horizontal* lines divide the abdomen into three zones, upper, middle, and lower.   The *upper* line encircles the body through the most prominent of the lower *costal cartilages, the ninth.*   The *lower* line encircles the body through the *most prominent part of the iliac crest.*   Two *vertical* lines subdivide these zones into regions.   They ascend from the centre of *Poupart's ligament* to the *costal cartilage* (the eighth).

The *nine* regions thus defined are :—

Right hypochondriac	Epigastric	Left hypochondriac.
Right lumbar	Umbilical	Left lumbar.
Right iliac	Hypogastric	Left iliac.

The viscera lie in those regions as follows : The *stomach* lies in the left hypochondriac, epigastric, and part of the right hypochondriac regions. But this viscus is capable of much alteration of position. When distended it may push up the diaphragm and encroach on the space allotted to the heart, as high as the fourth rib, or even into the axillary region. It may be pushed up by accumulation of fluid, or by tumour in the belly. It may be pushed downwards by fluid in the pleura, or by compression of the hypochondria with tight lacing. It may also be dragged downwards by the great omentum entering a hernial sac. The right end being less fixed undergoes more displacement than the left.

The *pylorus* can be felt when hardened and enlarged by disease, on the right of the umbilicus.

The *liver* is placed in the right hypochondriac, the epigastric, and to a small amount in the left hypochondriac regions. Its position in health changes when the diaphragm rises or falls, and with the position of the body. The lower border corresponds in the nipple line with the margin of the ribs, but when the body is erect, it falls below this line, and can be felt with the hand. When lying down the anterior border may disappear within the hypochondrium, though usually it is still perceptible. Behind the linea alba the anterior border reaches nearly half way to the umbilicus. The upper surface rises in ordinary respiration to the fifth rib, but forced respiration carries it as high as the fourth rib. On the surface of the body the upper surface of the liver lies about *one inch below the level of the nipple* at the *lower border of the pectoralis.* Percussion reveals its area by the dull note returned over the liver's surface, which reaches in the nipple line the upper border of the sixth rib. In the *back* the broad border is opposite the *twelfth dorsal* and *first lumbar* vetebræ.

The fundus of the *gall bladder* cannot be distinguished, except when distended or filled with gall stones. It is placed at the anterior margin of the liver *behind the outer border of the rectus, opposite the ninth cartilage.*

The large intestine begins in the right iliac region, and ascends through the right lumbar to the hypochondriac region. It then crosses the abdomen above the navel, through the umbilical or umbilical and epigastric regions, to the left hypochondrium. Finally it descends through the left lumbar and iliac regions to the pelvis, where it ends at the anus. In the left iliac region is the sigmoid flexure of the descending colon. Scybala in the gut may often be felt through the abdominal wall, their mobility usually suffices to distinguish them from tumours. Like the stomach, the great intestine is often much disturbed from its usual position.

In intussusception the invaginated part of the intestine forms a firm rounded tumour that may be distinctly felt along the course of the great intestine through the umbilical, left hypochondriac and lumbar regions towards the pelvis.

The *duodenum* is more firmly fixed than any other part of *the small intestine.* It courses round the head of the pancreas, opposite parts of the first three lumbar vetebræ, *from two inches above the umbilicus to the level of that landmark.* The pyloric end moves with the changes of position of the stomach. It occupies part of the right hypochondriac, right lumbar, and umbilical regions. The *ileum* and *jejunum* are placed in the umbilical, hypogastric, both lumbar and both iliac regions. Part often descends into the pelvis.

The *great omentum* is usually spread out over the small intestine in the umbilical, hypogastric, lumbar and iliac regions, but it may be tucked up into the left hypochondrium, or part of it may be dragged into some of the outlets of the body as a hernial protrusion.

The *pancreas* is behind the stomach in the left hypochondriac, umbilical and right lumbar regions. In very thin persons it may be felt *two inches above the umbilicus,* crossing the aorta opposite the twelfth dorsal and first lumbar spine.

The *spleen* lies in the left hypochondrium, *opposite the ninth, tenth, and eleventh ribs.* Its area of dulness, which corresponds to these three ribs in the infraaxillary region, is imperfect, as this viscus is overlapped above by the lung. In health its anterior margin does not project beyond the ribs. In disease it extends downwards, towards the umbilical region, and is readily felt. The notched anterior border will often serve to distinguish an enlarged spleen from other tumours in this situation.

The *kidneys* lie in the lumbar regions, on the psoas and quadratus, and on the diaphragm and last rib, being placed above the level of the umbilicus, nearly two inches above the iliac crest on a level with the eleventh and twelfth dorsal spines and first and second lumbar spines. The right is lower than the left kidney. They are very seldom felt when in the ordinary condition. To seek for the kidney, place one hand near the outer border of the rectus below the ninth cartilage, and the other behind on the erector spinæ. Then pressing them firmly together, bring both hands slowly outwards. If the kidney is enlarged, it will be felt indistinctly, as the hands reach the border of the erector. When the kidney is moveable it will be felt changing positions under the hands, and pain will be caused to the patient. When the kidney is inflamed, this manipulation excites aching pain. Besides enlarged kidney, peri-nephritic abscess may be detected in this region. Lower down, in the iliac fossa, psoas and iliac abscess may be felt ; which have to be diagnosed from peri-nephritic abscess.

The *abdominal aorta* passes down the abdomen a little to the left of the mesial line of the body, and usually *divides* at the level of the *highest part of the crista ilii,* or opposite a point on the surface *one finger's breadth below and one to the left* of the umbilicus. The point of division is, however, sometimes exactly in the mesial line, and then corresponds to a point immediately below the umbilicus. It may be compressed against the body of the third lumbar vetebra at the level of the umbilicus. Above the navel, pressure would injure the pancreas or solar plexus, and the vessel is usually compressed near its bifurcation.

The *cœliac axis* arises in the epigastric region, opposite the twelfth dorsal spine and between four and five inches above the navel.

The *superior mesenteric artery* arises in the umbilical region, *above the level of the umbilicus*, behind the pancreas, opposite the body of the second lumbar vertebra.

The *solar plexus*, surrounding the cœliac axis and superior mesenteric vessels, is in the epigastric and umbilical regions, opposite the two upper lumbar vertebræ, *at the level of and above the umbilicus.*

The *common iliac* arteries, about two inches long, course from the bifurcation of the aorta at the level of the highest part of the iliac crest along a line slightly arched outwards towards *a point midway between the anterior superior iliac spine and the symphysis pubis.* They divide again about the *level of the anterior superior iliac spine ;* (*exactly* opposite the intervertebral substance between the last lumbar and first sacral vertebræ.)

The *external iliac* continues this course to Poupart's ligament, below which arch it becomes the common femoral vessel.

Poupart's ligament, the symphysis pubis, the anterior iliac spine and iliac crest, are thus the surface marks for exposing the iliac arteries.

Inguinal Hernia.—The landmarks useful in examining the groin for hernia are the *pubic spine, anterior iliac spine, Poupart's ligament, external abdominal ring, spermatic cord, vas deferens, and testis.*

The *pubic spine* is at the level of the *great trochanter*, a point always easily detected, and the finger will reach the spine if carried directly inwards across the groin from this point, when, as in women, it is not convenient to find the spine by pushing the finger beneath the mons veneris.

The *external abdominal ring* lies just above and outside the spine. Generally both pillars can be felt, and always the outer one, which is attached to the spine. Besides giving passage to the cord, it is usually wide enough to admit the-tip of the little finger in health. When dilated by the passage of hernia, the finger can pass within the canal as far as the *internal abdominal ring*. Along this canal a hernia, or a fluid tumour, or enlarged spermatic cord may be felt easily in thin persons. The *internal abdominal ring* is ½ or ⅔ inch above Poupart's ligament, and is found by drawing a line at right angles to the ligament from a point midway along it, or vertically upwards from the position of the external iliac artery. The ring is placed about ⅔ inch above Poupart's ligament. In old herniæ the weight of the protrusion has often dragged the internal ring opposite the external one, and a single opening, wide enough to admit three or four fingers, is formed, through which the posterior surface of the pubes and a part of the pelvic cavity may be felt.

The *epigastric artery* cannot be distinguished on the surface, but it runs from midway between the anterior superior iliac spine and the symphysis pubis along the *inner side* of the *internal ring*, to the rectus, and gains the sheath of that muscle pretty *nearly midway between the pubes and the umbilicus.* When the internal ring is drawn towards the middle line, the course of the artery becomes more vertical.

The *contents of the scrotum* must be examined in diagnosing the nature of a tumour of the groin or scrotum. The *spermatic cord* with the *vas deferens* at the hinder part feeling like a whip-cord, the *testis*, the *epididymis* with the *globus major* above, the *globus minor* below, can all be distinguished in the natural condition of the parts. Sometimes the epididymis descends on the front or outer part of the testis, instead of at its regular position; in such cases it is much more loosely attached to the testicle.

The testis in scrotal hernia lies at the bottom of the scrotum behind the protrusion, and is often concealed by the hernia.

Inguinal hernia is readily distinguished from nearly all tumours connected with the testis by the latter not being continued along the inguinal canal.

Certain affections of the testis cause swelling along the canal. They are—*a*, incomplete descent of the testis; *b*, fluid in the sheath of the cord; *c*, varicocele; *d*, malignant disease of testicle invading the cord; *e*, abscess pointing along the cord; *f*, glandular tumours of the canal. Poupart's ligament passing *below the neck* of the mass to the spine of the pubes distinguishes the inguinal from femoral hernia and tumours of Scarpa's triangle proper.

Femoral Hernia has the following landmarks :— The *pubic spine*, the *anterior iliac spine*, *Poupart's ligament*, and the *femoral artery*.

The *saphænous opening* is about 1½ inch external

to the pubic spine, on a level with or a little above it.

The *falciform border*, when stretched by carrying the thigh outwards, can be distinguished. If the thigh is flexed and adducted, the fascia becomes lax, and the rupture, generally overlying Poupart's ligament, can be made out as a continuous mass passing inwards between the femoral artery and the pubic spine, at the deepest point of which tumour an impulse on coughing can be usually perceived.

Besides hernia, Scarpa's space may be the seat of enlarged lymphatic glands, aneurisms, fatty and cystic tumours, psoas abscess, and enlarged bursa beneath the psoas tendon.

In addition to the distinctions derived from the different relation of inguinal and femoral hernia to the more prominent landmarks, *inguinal* hernia is usually pear-shaped, and lies *above* the fold of the groin. *Femoral* hernia is more or less globular to the touch, and lies commonly *below* the fold of the groin. The *ring* that is *not* occupied by the rupture is almost always easily made out.

In employing taxis to reduce an *inguinal* rupture, bend the thigh and grasp the largest part of the mass with one hand and compress steadily, while the fingers of the other hand move the neck gently backwards and forwards to pass in the parts nearest to the abdomen. For *femoral rupture*, the thigh must be adducted as well as flexed, and the mass of the tumour drawn down a little before it is pressed steadily upwards to the crural ring.

Two chains of lymphatic glands are placed in the

groin. One lying along Poupart's ligament is affected
by irritation of the penis and scrotum or labium, of
the parts about the anus, and by strains of the abdominal
wall. The other, more vertical, overlying the femoral
vessels, is affected by irritants acting on the lower
extremity. Some of the group can be felt under the
skin in thin persons in their healthy condition ; chronic
irritation enlarges those naturally large enough to be
felt, and enables a much greater number of small ⁰nes
to be detected.

The distribution to the glands of the absorbent
ducts as they arrive from the several regions varies
much. For example, all the absorbents of the penis
may pour their contents into the glands of one side
only of the body : and the particular gland to receive
the absorbed fluid is not always the nearest. Most
commonly this is so, but in some bodies the afferent
vessel passes the nearest to open into a neighbouring or
even somewhat distant gland. This fact must not be
overlooked when tracing the cause of enlarged or
inflamed lymphatic glands.

The Perinæum.

This region is most easily examined when the body
is in the position for lithotomy.

The *limits* of the perinæum are readily made out
on the surface. At the sides are the *tubera ischii*
opposite, and about two inches from, the anus. From
them the rami converge forwards to the *symphysis
pubis.* In the middle line behind is the *coccyx.* Between the coccyx and the tubera ischii the firm flesh
is supported by the *sacro-sciatic ligaments.* Within

these boundaries the surface is arched and moderately firm in front of the anus, but soft and yielding behind that outlet.   The *raphé* divides the area longitudinally, and is the guide to the urethra and for perineal incisions.   An imaginary line from the front of one tuber ischii to the other divides the *urethral triangle* from the *anal triangle*.   The *central point* of the perinæum is nearly one inch in front of the anus. Just anterior to this point, the bulb of the urethra, with the corpus spongiosum passing forwards, is to be felt in all but very fat persons and young children, in the latter of whom the bulb is very small.   In a *thin* person the lower border of the triangular ligament can be felt just below and behind the bulb.   The front of the bulb is the posterior limit of the scrotum, a common place for perineal abscess and for urinary fistula to point. On each side of the ramus of the pubes the crus penis is distinct.

The urethra *passes through* the triangular ligament *one inch below* the symphysis pubis, and $\frac{3}{4}$ *inch above* the central point of the perinæum.

The bladder is generally about three inches from the surface, but this depth varies greatly in different persons.

At the *anus* in health the outlet shows brownish skin drawn into radiating folds.   If the anus be greatly opened, the mucous membrane becomes visible.   A palish line marks the junction of the skin and mucous membrane which corresponds to the lower margin of the internal sphincter.

The margin of the anus contains a large number of follicles, where subcutaneous abscess may form and be mistaken for fistula.

The anus may be closed by membrane—*atresia ani.*

To examine the rectum, let the patient lie on one side with the knees drawn up that a good light may fall on the anus. When the prostate is specially to be examined, the patient should lie on his back with his knees well bent. The finger, guarded by a layer of soap under the nail and round the margin, and well oiled, should be very slowly introduced to avoid causing pain and spasm. While the finger is travelling inwards, the tip should be applied to the mucous surface on all sides to seek for fissures, ulcers, &c.

The finger passed within the anus detects—

1. The sphincter ani and its upper border. When the finger is tightly grasped, there is probably a fissure or ulcer of the mucous membrane on the sphincter, generally just at the entry, and often over the tip of the coccyx. The internal opening of a fistula is rarely far above the upper border of the sphincter ani.

2. The prostate lies $1\frac{1}{2}$ inch within the anus, *i.e.* just past the sphincter. Its characteristic shape and firm consistence should be noted. When not enlarged, the finger passes beyond its base. When inflamed, it is very sensitive, and doughy to the touch. Abscess may be discovered by fluctuation over a limited area.

3. If a catheter be in the urethra, the membranous part of the urethra can be traced.

4. The trigone or base of the bladder. When the bladder is greatly distended, the trigone and bas fond fill up the rectum as a soft yielding tumour which fluctuates when the apex is palpated above the pubes. Usually the peritoneal lining of the back of the bladder and front of the rectum does not pass between the

trigone and the piece of the gut which lies behind it, thus the rectum is uncovered by peritoneum for a distance of four inches from the anus. In such cases the operation of tapping the bladder at the trigone is safely performed through the rectum. Occasionally the recto-vesical pouch reaches the prostate, and would be inevitably punctured twice by the thrust of the trocar through the rectum into the bladder.

5. The vesiculæ seminales can rarely be made out, unless affected by tubercular disease; this usually occurs in cases of tubercular testicle.

6. Higher up than the preceding are the transverse folds of the mucous membrane of the rectum, and in health its velvety softness may be felt. This part of the rectum is very loose and easily torn in old people, so that if the clyster pipe be roughly introduced, it may perforate the wall of the rectum and the clyster be thrown into the ischio-rectal fossa, or even into the cavity of the peritoneum. To avoid this, use great gentleness and direct the nozzle backwards.

7. Fæcal accumulations.

8. Stricture; that from syphilis usually begins at the anus, with much induration of the skin and mucous membrane surrounding the outlet; that from malignant disease commences three or four inches up the canal.

9. Hæmorrhoids depending from the last two inches of the gut; or polypi.

10. Abnormal development of the rectum, imperforate rectum, communication between the rectum and the vagina.

11. Ischio-rectal abscess.

12. In children a stone, or the sound in the bladder.

13. Also in children, the line of attachment of the recto-vesical fascia, the sciatic notches and ligaments, the coccyx and concavity of the pelvis, and even the brim of the pelvis.

14. Ovarian and uterine and other pelvic tumours.

The Upper Extremity.

The *surface marks of the shoulder* are of much importance in ascertaining the seat and nature of injuries and of diseases which are met with about that joint. Several of them are also guides for the direction of incisions during operations.

The following landmarks may always be detected. The *clavicle* in its whole length, the *spine and acromion* process of the scapula, the *great tuberosity, the shaft* of the humerus ; and if the fingers are passed into the arm-pit, the *neck* and *head* of that bone. The deltoid, and coraco-brachialis muscles with the border and tendon of the great pectoral muscle are more or less distinct.

Generally also can be distinguished the *inferior angle, inferior* or outer and *inner* or posterior borders cf the scapula, with the *coracoid* process placed deeply in the depression between the great pectoral and deltoid muscles (*infra clavicular fossa*). The relative changes of position which these landmarks undergo, are also important in diagnosing the nature of the injury or disease under examination. When the arm hangs supine ("little finger to the seam of the trousers"), the acromion, the external condyle of the humerus and the styloid process of the radius are in the same line on the outside, while the head and the internal condyle of the humerus are in a line with the styloid process of the ulna. In this position the *bicipital groove*, along which abscess causes obscure swelling, looks *directly forwards*.

In the natural condition, a *projection* is sometimes developed at the acromial end of the *clavicle*, which may be mistaken for dislocation upwards of that bone when swelling follows a severe contusion of this part. Again, the *symphysis* between the *acromion* and the spine remains in rare instances mobile throughout life, and may suggest a fracture of the acromion. If these conditions are natural they will exist on · both shoulders. Hence, in examining a limb suspected of injury, it is always well to examine the corresponding part of the opposite side.

The characteristic *roundness of the shoulder* is due to the deltoid muscle and the upper end of the humerus. *It is lost* in dislocation or in atrophy of the muscle from paralysis, when the head falls away from the acromion, and the aspect of the shoulder is much changed. On the other hand the growth of tumours, fractures of the neck of the humerus, effusions into the capsule or neighbouring synovial bursæ, more or less *increase* the fulness of the shoulder. Fractures of the anatomical neck cause little alteration to the shape of the shoulder. If the bone is broken at the surgical neck, the axis of the shaft is drawn towards the coracoid process by the pectoral muscle; the front of the shoulder is fuller than natural.

In deciding whether the upper end is broken or is

separated at the epiphysis, it must be borne in mind that the head and tuberosity unite in the *fifth* year, and the shaft and upper end in the *twentieth* year.

To define the tuberosities and head of the humerus, the surgeon should rotate the arm with one hand, holding the elbow bent, while the other is placed over the acromion and deltoid. A larger surface of the head may be felt by the fingers placed in the top of the armpit during the rotation of the arm.

In measuring the humerus, carry the tape from the *symphysis*, not from the extremity of the acromion to the outer condyle.

Along the *anterior border* of the deltoid is a groove. In this lies the cephalic vein before it dips between the deltoid and pectoral muscles to reach the axillary vein.

This *groove in front of the deltoid* also marks the position of the coracoid process; immediately within and below that, is the *axillary artery* with the vein to the inner side and the cords of the brachial plexus to the outer side. The nerves afterwards surround the artery and the course of the nervo-arterial cord can be felt in the axilla from the *first rib to the teres tendon.* There the artery becomes the brachial trunk. The vessel lies against the neck of the humerus in the axilla, if the arm is drawn away from the body, and may be compressed against that bone.

In *cutting down* upon this artery the surface guide is the *inner border of the coraco-brachialis* muscle.

The *posterior border* of the shoulder is formed by that of the deltoid, and is usually easily defined. *Beneath* the deltoid muscle, protecting it from the upper end of the humerus, is a large *bursa ;* effusion into this bursa follows blows on the shoulder, and must be distinguished from effusion into the capsule of the joint. A bursa is also interposed between the coracoid process and the tendon of the subscapularis close to shoulder joint with which it communicates. This bursa lying near the neck of the scapula may swell and simulate an enlarged gland or abscess at the apex of the axilla.

In the *arm* the main components are easily defined. Behind is the triceps rounded above and flattened at the elbow into the tendon which is attached to the olecranon. In front, the biceps contracts suddenly just above the elbow into its tendon which is passing to the tubercle on the radius. *In the groove on the inner side* are the vessels and nerves, having for their guide the *inner border of the biceps* which overhangs them in the middle of the arm. Besides the deeper vessels the basilic vein lies in this groove. It can be seen in the lower part of the arm until it dips beneath the fascia to join the brachial vessels.

The *humerus* may be felt below the attachment of the deltoid muscle on either side of the biceps muscle, where the supracondylar ridges are easily made out. The swelling contour of the arm on the outer side of the biceps is due in the upper part to the brachialis anticus, and in the lower part in front of the condyle to the supinator longus and extensor carpi radialis.

The artery may be compressed against the bone throughout its course.

The Elbow.—The *bony land-marks* are, on *the outer*

side, the *external condyle* with the external supra-condylar ridge rising from it. Below and a little behind the condyle is a *depression*, well marked in most persons, in which the *head of the radius* can be felt rotating during pronation and supination of the fore-arm. On the inner side is the *inner condyle*, so pro-minent under the skin that it needs careful padding whenever an L-shaped splint is worn inside the arm. The inner supracondylar ridge can also be plainly felt. At the back is the olecranon ; subcutaneous at its posterior surface, but at its upper end covered by the insertion of the triceps muscle ; between this pro-minence and the inner condyle is the *cord of the ulnar nerve* or " funny bone." It can sometimes be dis-tinguished by the examiner ; but a blow on it pro-duces the well-known painful sensation to the sufferer.

The normal relation of the bony points at the elbow has to be carefully studied. The *olecranon* in *extreme flexion* of the elbow is *in front* of the condyles. In rectangular or semi-flexion, the olecranon is *immediately beneath* the condyles. In extreme extension the ole-cranon *is on a level with* and behind the condyles. In *fracture of the humerus just above* the condyles, the relations of these landmarks are unchanged, though the position of the condyles and olecranon to the shaft of the humerus is much altered. The elbow is carried backwards and the lower end of the shaft causes projection forwards of the belly of the biceps *above* the hollow in front of the elbow. This deformity is usually without much difficulty reduced with crepitus, but easily recurs if the limb is left to itself.

In *dislocation* the relation of the olecranon is greatly altered. That process is carried away from the lower end of the humerus, is more or less firmly fixed, and if carried backwards, the lower end of the humerus forms a hard *fulness obliterating the hollow* in front of the elbow. In this case also, unless the swelling be very abundant, the great sigmoid notch can be distinguished *behind* the condyles.

The *lower epiphysis* of the humerus joins the shaft in the *eighteenth* year. Before this age, disjunction may take place. In this case the signs are more obscure, the grating is not like bony crepitus, and the exact point where the humerus is divided is difficult to make out. But the maintenance of the natural position of the olecranon in regard to the condyles is still the chief distinction from dislocation.

When fluid is effused into the elbow-joint, swelling is produced on either side of the triceps insertion and just below the outer condyle, quickly filling up the pit under which the head of the radius rolls. Swelling over the tip of the olecranon is caused by effusion into the bursa between the skin and the bone. Swelling, obscure and very painful, at the back of the olecranon is caused by effusion into the bursa between the tendon of the triceps and the olecranon. Swelling situated in the neighbourhood of the inner condyle, accompanied by brawny infiltration of the skin and fluctuation, is sometimes caused by inflammation of the epitrochlear lymphatic gland which lies in front of the fascia and internal inter-muscular septum just above the condyle. There are occasionally two lymphatic glands; they receive absorbent vessels chiefly

from the inner side of the surface of the forearm and
little fingers, and hence are irritated most commonly
by inflammation of that part.

In the *forearm*, above the wrist, there are few salient
points. The ulna is subcutaneous from the olecranon
to the styloid process. The radius can be readily felt
in the lower third of the forearm. Anteriorly and
posteriorly the muscles make fleshy prominences in
the upper part. In the. lower part the tendons of
several can be distinguished, and serve as landmarks.
The contour of the muscles is affected to some extent by the attachment of their aponeurosis to the
ulna along the inner and posterior parts of the forearm.

The *hollow in front of the elbow.* Here the superficial
veins of the forearm receive large affluents from the
interior; their copious supply of blood renders them
suitable for venæsection.

The median vein of the forearm divides at the apex
of the hollow into the median cephalic on the outer
side, and median basilic on the inner side. The outer
branch joins the radial vein (or veins,) at the outer
side of the elbow, and, forming the *cephalic* vein, mounts
the outer side of the arm. The inner branch, that
which usually receives the deep affluents, passes inwards
in front of the condyle to join one or two ulnar veins
and form the basilic vein. As the median basilic inclines to the ulnar veins it overlies a tough band from
the biceps tendon to the fascia of the forearm, which
separates the vein from the brachial artery beneath.
Being thus more firmly supported and more copiously
supplied with blood, it is usually selected for venæ-

section. Too deep an insertion of the lancet may penetrate through the floor of the vein, the aponeurosis from the biceps, and the coat of the artery, and thus establish a communication between the artery and the vein, which if permanent, will form an aneurismal varix.

The brachial artery divides in the hollow one finger's breadth below its centre into the radial and ulnar arteries.

Occasionally a large artery passes over the inner part of the anterior aspect of the elbow just between the anterior and posterior ulnar veins. It is the ulnar artery taking high origin from the brachial. The possible presence of such a vessel should be remembered.

On the *posterior aspect of the wrist*, at the inner side, is the *head of the ulna*, the most prominent part of the bone. The part felt under the skin projects between the tendon of the *extensor carpi ulnaris* at the inner side, and the tendon of the *extensor minimi digiti* at the outer side. The tendon of the extensor carpi fills the groove in the head of the ulna, and can usually be distinguished from the bone. A little in front of the tendon of the extensor carpi is the *styloid* process. When the hand is *pronated*, the *head* of the ulna is most prominent and easily detected. In *supination*, on the contrary, the *styloid* process is most prominent, for the radius in supinating round the ulna, raises the extensor muscles above the level of the head in that position. The *cuneiform bone* can be felt at the back of the wrist below the head of the ulna, overlaid by the tendons of the ulnar extensor and extensors of the little finger.

At the *radial* side of the back of the wrist is the lower extremity of the radius, overlaid by tendons of the extensors. The *extensor secundi internodii pollicis* becomes prominent when the thumb is extended, and is then the guide to the oblique groove on the posterior margin of the radius, through which it plays. At this groove the radius is frequently broken in Colles' fracture. The *styloid process* of the *radius* is overlaid by the tendons of the two first extensors of the thumb, which somewhat conceal it. Immediately below the styloid process is a prominence formed by the *tubercle of the scaphoid* and the *trapezium.*

The styloid process of the radius reaches a little lower in the limb than the styloid of the ulna.

When the thumb is extended, the tendons of its extensor muscles form two prominent ridges. In the hollow between these ridges, at the back of the carpus, just below the end of the radius, are the *radial vein* and the two tendons of the radial extensors of the wrist; more deeply still lies the last part of the *radial* artery.

The superficial veins can be seen coursing over the back of the hand and wrist, while towards the knuckles, the extensors of the fingers are also distinguishable when they are set in action. If the ring finger be moved while the other fingers are rigidly extended, the connecting bands from it to the tendons of the middle and little fingers, can be seen or felt sliding under the skin, near the knuckle.

The *posterior annular ligament* cannot be defined beneath the skin ; but the bones to which it is attached can be readily distinguished, viz., at the outer end, the back of the lower end of the radius, near the styloid process, and at the inner end the back of the cuneiform and pisiform bones. It is divided into six compartments for the tendons : four are on the radius. The *outermost* for the two first extensors of the thumb ; the *second* for the two radial extensors of the wrist ; the *third* oblique one for the extensor secundi internodii ; the *fourth* gives passage to the common extensor of the fingers and the extensor indicis ; the *fifth* lies between the radius and ulna, for the extensor minimi digiti ; and the *sixth* is attached to the inner part of the head of the ulna, just external to the styloid process to give passage to the extensor ulnaris.

Bursal tumours, or " ganglia," are frequently found at the back of the wrist and hand, in the synovial sheaths.

At the front of the forearm, the *radius* can be felt in all persons, and the *tendon of the flexor carpi radialis* can be felt or seen in the lower part, as far as the wrist, where it disappears under the annular ligament opposite the styloid process of the radius. This tendon is the guide to the *radial artery*, when that vessel is obscured by swelling. In the natural state the artery can be felt or seen to pulsate in front of the lower end of the radius on the *outer side* of this tendon. The *position* of the artery in the forearm is denoted on the surface by a line drawn from the *centre of the hollow in front of the elbow*, to the *styloid process* of the radius. Sometimes the *superficial volar* branch rises two or three inches above the wrist ; it then accompanies the radial trunk, and thus is produced the double pulse (" Pulsus duplex ").

At the *centre of the forearm*, about the wrist, unless the patient be very fat, or the muscle be absent, the tendon of the *palmaris longus* can be felt. It passes over the annular ligament, and is the surface guide for the *median nerve* which lies close to its inner side.

At the *inner side* of the forearm the tendon of the *flexor carpi ulnaris* is felt in all but the very fat, and if traced downwards, the attachment to the pisiform bone below the styloid process of the ulna can be distinguished. This muscle is the guide for the *ulnar artery*, which lies close to the outer side of the tendon. The tendon usually overhangs the vessel; though at the wrist the artery becomes quite superficial before it enters the palm, over the annular ligament, close to the pisiform bone. The *ulnar* nerve is on the inner side of the artery.

The line of the radio-carpal articulation crosses the front of the wrist about ¾ inch above the transverse crease in the skin which separates the wrist from the palm.

The *anterior annular ligament*, like the posterior, cannot be defined in the healthy condition of the wrist. But when the synovial bursa of the flexors of the fingers is distended with fluid, its limits often become distinct. It is attached to the pisiform and cuneiform bone at the inner end, and to the scaphoid and trapezium at the outer end. The tendon of the flexor carpi radialis passes in a separate synovial sheath through the attachment of the outer end of the ligament. Sometimes distension of this sheath forms a small bursal tumour, or "ganglion," just above the annular ligament, close to the styloid process.

The *large synovial sheath,* for all the tendons of the two flexors of the fingers, reaches underneath the annular ligament as far *as the middle of the palm,* and *above the wrist for* 1½ *or* 2 *inches.*

To open it or to *open an abscess of the front of the forearm,* near the wrist, an incision may be safely made *close* to the *inner* side of the flexor carpi radialis; at this point the *median nerve,* in the *centre* of the limb, and the *radial artery* at the *outer border* of the tendon will be avoided.

In *the palm* two transverse creases in the skin correspond to the level of the articulation of the metacarpal bones with the phalanges.

The prominent fleshy mass on the outer part of the palm is called the ball of the thumb, or *thenar prominence ;* that on the inner part, the *hypothenar prominence.*

In the hollow between the prominences the fascia is very thick, and beneath it the *superficial* palmar *arch* of the ulnar artery courses opposite a *line drawn* on the surface, with a slight downward convexity from the pisiform bone to the *web of the thumb.* This is the vessel that is usually injured in *"wounds of the palmar arch."* The *deep palmar arch* of the radial artery lies beneath the tendons on the bases of the metacarpal bones, and a little nearer to the wrist than the superficial arch, nearly opposite to the centres of the thenar and hypothenar prominences on the surface.

The *digital branches* of the superficial arch arise *opposite the clefts of the fingers,* so that *abscess of the palm* must be opened opposite the *centre* of the fingers, over the head of the metacarpal bone, to avoid the vessels and nerves, and in the *lower half* of the palm, to avoid the superficial arch and the common sheath of the tendons under the annular ligament.

On the fingers the knife should be kept *to the middle* of the back or front, that the digital vessels may not be wounded when laying open a thecal abscess or whitlow.

In amputating a phalanx, it should be recollected that the articulation lies beneath the *lower* of the two creases, into which the skin falls at the *back* of each joint, and that the projection in each side of the base of the phalanx can be felt a little *above* the crease on the *palmar* surface of the joint.

The Lower Extremity.'

The *surface* marks of the *groin, hip, and buttock* are *anteriorly*, the swell of the tensor vaginæ femoris, and the sartorius on the outer side, and the swell of the adductors on the inner side ; between them is the hollow of Scarpa's space. *Posteriorly*, the swell of the gluteus maximus, abruptly limited below by the *gluteal fold*.

The *bony* landmarks are—at the *upper* and *posterior* part, the iliac crest, from the anterior spine to the posterior spine. In the middle line behind, at the bottom of a deep groove, between the glutei, are the spines of the sacrum and coccyx. *At the inner and lower part* are first, the pubes, including the symphysis, spine, horizontal and descending rami; then the ramus and tuberosity of the ischium. *Externally* is the *great trochanter*, easily recognised in all persons, its situation being at the bottom of a depression when the femur is everted, and prominent when the limb is rotated inwards or adducted. It projects further laterally than the iliac crest. The tip of the great trochanter is ¾ inch below the level of the head of the femur, and on a level with the pubic spine. Hence the horizontal furrow across the front of the thigh between these processes passes over the capsule of the hip joint. Fulness here, in the line of the artery, with tenderness, are present when the capsule is distended with fluid, as in early stages of hip disease.

Deep abscesses point in Scarpa's triangle. Those coming forward *below* Poupart's ligament, and pointing *internally* to the artery are usually psoas. A psoas abscess enters the thigh, like the muscle the sheath of which it occupies, outside the vessels, and has passed behind them if it point on the inner side of the thigh. When still *below* Poupart's ligament, but pointing *external* to the artery, the abscess comes generally from the hip joint. If external to the artery, but above Poupart's ligament, it is probably iliac.

The *great trochanter* is altered in its relations to the crista ilii and anterior spine, by dislocation of the head of the femur, by fracture of the neck, and by absorption of the head and neck of the bone in morbus coxæ.

In *dislocation* to the dorsum, the trochanter is carried further from the spine than natural, but is nearer to the crest. In the other dislocations it is approximated to the spine in pubic, but carried downwards from both crest and spine in thyroid dislocation.

Nelaton's line.—According to Nelaton's observations,

in health, the upper border of the great trochanter is crossed by a line from the anterior superior iliac spine to the most prominent part of the tuber ischii, in every position of the thigh. When there is dislocation backwards, the trochanter has passed above this line, towards the spinal column. In fracture of the neck this line will pass above the great trochanter.

When the shaft of the uninjured femur is rotated, it traverses the arc of a circle, having the head and neck for a radius ; but when the neck is broken, the trochanter revolves in the axis of the shaft, and consequently through the arc of a much smaller circle.

Mr. Bryant employs this changed relation as a pathognomonic sign of fracture of the neck. He places the body in an exactly horizontal position, and drops a vertical line from the anterior spine. When the neck of the femur is unbroken, the distance between the tip of the trochanter and this line is the same on both sides of the body. When there is fracture, the distance is shortened in amount corresponding to the shortening of the broken neck.

Wasting of the gluteus, and consequent *loss of depth to the gluteal furrow* are early signs of hip disease.

This *gluteal furrow* is the best place to feel the *great sciatic nerve* at the level of the ischial tuberosity. It is midway between that point and the great trochanter. Pressure here produces pain if the nerve is affected. Pressure higher up on the gluteus, opposite the tip of the great trochanter, *may* give pain by compressing the capsule of the hip-joint when that is inflamed.

The *gluteal artery* issues from the pelvis at the *juncture* of the upper and middle thirds of a *line drawn from the posterior superior iliac spine to the great trochanter* when the thigh is rotated inwards.

The *ischiatic artery* leaves the pelvis about *half an inch* below the gluteal on the same line.

The *pudic* artery lies over the ischial spine, at the juncture of the lower and middle thirds of a *line drawn from the outer side of the tuber ischii to the anterior superior iliac spine*, when the thigh is rotated inwards.

If the body rests in a sitting posture on a firm level surface, the pressure is borne on the bony points only; if, however, the body is placed on a soft yielding cushion, some of the pressure reaches the blood-vessels just mentioned, and, interfering with their circulation, over-loads the inferior hæmorrhoidal, uterine and other vessels.

The *femoral artery* enters the thigh at the lower border of the horizontal ramus of the pubes, midway between the anterior iliac spine and the symphysis pubis. When the knee is half-bent and the thigh rotated outwards, the course of the vessel is denoted by a line from the point mentioned above *to the most prominent part of the tuberosity on the inner condyle of the femur*. At first the vessel is some distance in front of the neck of the femur, but at the lower part of Scarpa's space it lies along the inner side of the shaft, and should be there compressed by a tourniquet.

The most convenient place for *immediately* controlling the circulation through the lower extremity is the last part of the *external* iliac resting on the pubes, where the vessel can be felt to pulsate in all persons. The least fatiguing mode of applying pressure is to grasp the top of the thigh with both hands, the fingers passing backwards on the outside and inside, while the thumbs are thrust down on the vessel, one over the other. Pressure is kept up by the thumbs alternately to prevent the hands becoming fatigued by both being used continuously.

In the *region of the knee*, the following *bony* points can be distinguished :—

The *patella*, with its ligament, passing down to the *anterior tuberosity* of the tibia in front; the *tuberosity* on each *condyle of the femur;* the *external* and *internal tuberosities of the tibia :* the *head* of the *fibula ;* and midway between the head of the fibula and the anterior tuberosity, a small *eminence* on the tibia.

The *surface marks* vary with the position of the limb. *In front,* if the leg be extended, the patella and tendon of the quadriceps are in bold relief, with a furrow on each 'side of the tendon. *If the knee is bent,* the patella sinks into the trochlear groove, when the condyles with the space between them become very evident. The *tuberosities of the tibia* are more easily made out. In this position the tibia can be *rotated slightly on the femur.*

Externally, there is a depression which corresponds with the junction of the femur and tibia and external semilunar cartilage ; *above* this depression is the *external condyle of the femur; below* it, the *outer tuberosity of the tibia;* which is the landmark of the *line for entering the joint* in *resection.* Just behind the external *tuberosity are the head of the fibula* and *tendon of the biceps.* Below the highest part of the head of the fibula, behind and internal to the tendon of the biceps, is the *external popliteal nerve*. This may be divided in tenotomy of the biceps if the knife be carried too near the head of the fibula and directed inwards instead of outwards against the tendon.

Internally, between the internal condyle and internal tuberosity, the finger can be laid on the internal semilunar cartilage, a point sensitive to pressure in early stages of inflammation of the articulation from jar, or from strain of the lateral ligament. The point of attachment of the *tendon of the adductor magnus* muscle to the tuberosity on the internal condyle is on the level of the *line of the epiphysis*, and almost on *that of the upper border of the trochlear surface*. The *internal saphenous vein* winds round the limb behind this condyle.

When *effusion* takes place *into the capsule* of the joint, the furrows on each side of the ligament are filled to bulging, and the prominence of the patella and its ligament are lost, the former " floating " on its trochlea. The capsule ascends for *three fingers' breadth above* the patella, and its limit becomes plainly marked when distended by fluid.

When the knee is bent, the capsule is drawn downwards; hence, when operating on the thigh near the knee, *bend that joint*.

Effusion into the bursa patellæ pushes forward the skin over the lower part of the patella and the ligament into a prominent well-defined tumour which does not fill up the furrows at the sides of the ligament.

The *bursa between the ligamentum patellæ and the tibia* is occasionally distended with fluid. It makes no defined swelling, but a general fulness of the front of

the tibia partially occupying the lateral furrows. It is
usually much more painful than distension of the
superficial bursa.

Posteriorly, when the knee is straightened, the surface of the ham is convex, and the boundaries ill-defined,
the tendon least concealed is that of the *semitendinosus.*

The boundaries of the *popliteal* space are most
salient when the knee is somewhat bent. The *biceps*
on the outer side and the *semitendinosus* and semimembranosus on the inner, being easily distinguished in all
persons. The gracilis can sometimes be descried a
little further forward than the other two inner ham-
strings. The ilio-tibial band can be felt as a strong
cord beneath the skin, running down in front of the
biceps tendon to the external tuberosity of the tibia.
The bent position relaxes the fascia covering the
space, when the contents can be better examined. At
the *lower part* the *external saphenous vein* disappears to
enter the ham; this is very evident when the vein is
varicose. The *internal popliteal nerve* is placed in the
middle of the space just beneath the fascia, and when
the limb is nearly straight, the nerves can be felt
through the skin in some persons as round cords at
the upper end of the space.

The *popliteal* artery lying deeply at the bottom of
the space cannot be felt to pulsate readily. When its
pulsations are obvious, some morbid swelling conveys
the vibration to the surface. The vessel divides at the
lower border of the popliteus muscle into anterior and
posterior tibial arteries. This division corresponds to
a point on the outside of the leg, about 1¼ *inches* below
the head of the fibula.

The *tumours* which may develop in the popliteal space are, enlarged bursæ of the hamstring tendons, enlarged lymphatic glands, abscess connected with these structures or with disease of bone, exostoses, cystic and solid tumours and aneurism.

The *bursa* most commonly enlarged is that between the *semimembranosus* and the *inner head of the gastrocnemius*. When the knee is flexed it disappears or becomes flaccid, when the knee is straightened and the weight of the body thrown on the limb, it becomes a prominent, firm, somewhat irregular tumour in the middle of the ham. It does not throb with the artery. In about one of every five cases dissected, the bursa is found to communicate with the synovial capsule of the knee-joint.

The bursa between the *inner head* of the gastrocnemius and *inner condyle*, by distension makes a swelling behind that point of bone, ill-defined, but causing pain in walking. It frequently communicates with the joint; more often still, the bursa beneath the *outer head* of the gastrocnemius and the *outer condyle* is a prolongation of the joint's synovial capsule.

There is also a bursa between the semimembranosus and the tibia, and one between the tendon of the biceps and the external lateral ligament.

Abscess is usually connected with the lymphatic glands of the space, most of which are arranged deeply around the artery. When of slow formation, with moderate inflammation of the cellular tissue, it may vibrate with the arterial pulsation. It has *no expansile* thrill, and can often be sufficiently isolated from the vessel to lose its vibration.

Aneurism is distinguished by the signs proper to all aneurismal swellings, and generally by its limits being tolerably easily defined. Diffused or suppurating aneurism, though ill-defined, has special characters— among them, the state of the pulse in the arteries of the leg.

Solid tumours are more difficult to distinguish from aneurism if that be itself solidified or almost so. Then the several signs proper to tumours originating independently of the artery must be sought on the one hand, and the absence of those proper to aneurism ascertained on the other.

In the leg the bony landmarks are ; of the tibia, the *inner surface* for the whole of its length, the *inner tuberosity* and the *internal malleolus*, the anterior or *shin ridge* and the *inner border*. On the *outer side* are the *head of the fibula* above, and the *lower fourth of the shaft terminating in the external malleolus*, below.

In front the muscles form a smooth convex surface above, closely confined by the aponeurosis attached to the tibia and fibula. Below they divide into tendons that cross the front of the ankle-joint in three conspicuous divisions. The most internal, the tibialis anticus ; the next, the long extensor of the great toe ; both of these are on the tibia. The third, the bundle of the common extensor and peroneus tertius, is in front of the joint between the tibia and fibula.

On the *outer side of the calf* there is a groove, well marked when the muscles are in action ; this indicates the interval between the soleus and the peronei muscles. *Posteriorly*, the swell of the calf is caused above by the gastrocnemius, below by the soleus tapering into the tendo achillis at the small of the leg. Above the heel the tendon is easily distinguished in all persons.   It is narrowest opposite the ankle, the point where it is divided in tenotomy.

The internal saphenous vein crossing the internal malleolus in front, passes up the inner side of the leg. Varicosity of this vein renders it tortuous, prominent, and irregular.   It is accompanied by the internal saphenous nerve.

In a well-formed leg the *inner edge of the patella, the inner side of the ankle, and the inner side of the great toe are in a line*, thus forming a guide for the reposition of the fragments of a broken tibia.

The *anterior tibial artery*, rarely ligatured in its continuity, is sometimes to be secured at the bleeding point when injured.   The guide on the surface is a line drawn from the *inner side* of the head of the fibula to the centre of the *front of the ankle-joint*.   It lies mainly along the outer border of the tibialis anticus muscle.

The *posterior tibial* artery can be felt beating *about half an inch from the edge of the tibia, opposite the ankle*. The course of the vessel is represented on the surface by a line drawn from the *centre of the top of the calf* to *a point midway between the tip of the internal malleolus* and the inner tubercle of the heel.

The guides for the tendons cut in treating clubfoot are found as follow ;—

The tendon of the *tibialis posticus* lies close to the inner edge of the tibia, *midway between the posterior and anterior borders of the leg*, and may be made salient by abducting the foot, while the thumb is placed over the tendon. In children it is often very difficult to feel this tendon, and the midway point remains the chief guide. The point is *anterior* to the *artery*, which runs midway *between the tibia and the posterior border* of the leg at the ankle. The *peronei* tendons lie close *behind the fibula*, at the ankle, enclosed in a common sheath, and should be made tense by flexing the ankle and adducting it. The tendon of the larger muscle is most superficial, that of the peroneus brevis is close behind the bone. They are cut two inches above the tip of the malleolus.

The *tibialis anticus* crosses the front of the tibia and ankle-joint to gain the first metatarsal and internal cuneiform bone. It is usually both felt and seen without difficulty when the foot is rotated outwards. It is divided sometimes as it crosses the tibia above the ankle, but *most frequently on the scaphoid bone;* the knife being introduced between the artery and the tendon.

At *the ankle* the bony projections are :—the *internal malleolus* short and broad, is placed rather in front of the centre of the joint. The *external malleolus* is longer, more pointed and is opposite the centre of the joint. The *anterior margin* of the lower end of the tibia can be traced above the ankle-joint, crossed by the tendons of the extensors of the toes, and flexors of the ankle. They are the *landmarks* for the incision in Syme's amputation. Behind the outer malleolus the external saphenous vein ascends to the leg. The back of the ankle-joint is divided into two hollows by the *tendo achillis*, which is sub-cutaneous from its origin to its insertion at the heel.

The *Foot.*—In examining the foot the bony land-marks employed on the inner side of the foot are ; (1) the *tubercle* of the os calcis, (2) the *inner malleolus*, and (3), about one inch below it, the *sustentaculum tali*. In front of the inner malleolus is (4) the *tubercle of the scaphoid*. Besides these, there are (5) the *internal cuneiform* bone, and (6) the *base of the first metatarsal* bone ; its shaft and head.

Along the *outer* side of the foot are the following osseous points ; (1) The *outer tubercle* of the os calcis, (2) the *external malleolus*, (3) the *small tubercle on the os calcis between the peronei tendons*, the short tendon above, and the long tendon below it; (4) the pro-jecting *base of the fifth metatarsal bone.*

On the dorsum of the foot the scaphoid is prominent on the inner side, and the cuboid can be felt under the skin behind the base of the fifth metatarsal bone. *When the foot is extended* the head of the astragalus projects in front of the lower border of the tibia. The tendons of the extensors are easily distinguished along the dorsum of the foot to the toes.

The *dorsal* artery can be felt pulsating in a line from the centre of the ankle to the first interosseous space outside the base of the metatarsal bone of the great toe. The tendon of the long extensor of the great toe runs on the inner side of the vessel, and the inner tendon of the short extensor running on its outer side for most of its course, crosses the artery about half an inch before that vessel dips to the sole.

The *external plantar artery* traverses the foot obliquely from the *midway point* between the inner malleolus and inner tubercle of the os calcis to the *base of the fifth metatarsal* bone. In this part the thickness of the first layer of muscles intervenes between it and the plantar fascia, a structure it is necessary to divide in some forms of talipes.

From the *fifth metatarsal* bone to the *base of the first* metatarsal, the vessel is placed high in the sole close to the bones. From this part the digital arteries are given off *opposite* the webs of the toes. The *internal plantar* artery is separated from the surface by the thickness of the abductor pollicis. It is inconsiderable in size.

The *plantar nerves* have the same connection as the arteries.

The webs of the toes correspond nearly to the joint between the first and second phalanges.

The metatarso-phalangeal articulations are about one inch behind these webs.

For *Syme's* and *Pirogoff's amputations* at the ankle ; the external malleolus shows the centre of the joint, on the outer side of the ankle, and a point one-third of an inch behind the tip of the inner malleolus indicates the central point on the inner aspect.

The incisions start and terminate at these points.

For *Chopart's* amputation; on the inner side the joint between the scaphoid and astragalus is found just behind the *tubercle* of the scaphoid. On the outer side the joint between the calcaneum and cuboid bone is indicated by a point half an inch *behind* the base of the fifth metatarsal bone.

In *resecting* the first metatarsal bone the articulation between it and the internal cuneiform is one inch in front of the tubercle on the scaphoid.   The incision would begin *half an inch* in front of the scaphoid.

For resecting the fifth metatarsal bone, the incision is commenced opposite the tip of the base, half an inch in front of which joint lies the articulation.

LISTS

INSTRUMENTS AND APPLIANCES REQUISITE
OR OCCASIONALLY USEFUL

MOST OF THE IMPORTANT AND ORDINARY
OPERATIONS IN SURGERY.

PREPARATIONS AND REQUISITES FOR OPERATIONS IN GENERAL.

The Operating Room.
The Sick Bed and Bed-Room.

Sedatives and Restoratives.
The Arrest of Hæmorrhage.

OPERATING ROOM,

Having a good Light and Windows that open readily, and a Fire in Winter.

1. Firm table, 4 feet long, 2 feet wide, and 3 feet high.
2. Pillows.
3. Blankets.
4. Towels.
5. Old linen.
6. Macintosh sheets.
7. Old carpet, or old sheet to catch the blood.
8. Tray of sawdust or sand.
9. Bandages.
10. Strapping plaster.
11. Lint.
12. Oiled silk.
13. Cotton wool.
14. Tow.
15. Perchloride of iron.
16. Basins, large and small.
17. Hot and cold water, ice.
18. Bucket and slop-jar.
19. Sponges.
20. Chloroform and inhaler.
21. Oil.
22. Pins.
23. Scissors.
24. Brandy.
25. Ammonia.
26. Feeding cup.
27. Enema syringe.
28. Fire for heating cauteries.

SICK ROOM AND BED.

1. Iron bedstead.
2. Wool and hair mattresses.
3. Several pillows, soft and of different sizes.
4. Air and water cushions.
5. Blankets, small single ones.
6. Pieces of soft flannel.
7. Six sets of sheets and pillow cases.
8. Old soft linen.
9. Cotton-wool.
10. Towels.
11. Soft pocket-handkerchiefs.
12. Three pieces of Macintosh, 2 feet 6 inches square.
13. Bed cradle.
14. Light bedgowns.
15. Flannel jackets, and flannel Zouave drawers, or cotton jackets and pyjamas.
16. Hot-water bottles and bags.
17. Bed-urinal and bed-pan.
18. Night-stool.
19. Basins.
20. Cold water.
21. Condy's fluid.
22. Carbolic acid.
23. A bed-rest chair.
24. Night lights.
25. A fire, or in summer a lamp to burn in the fireplace, to create a draught of air.
26. Enamelled saucepan.
27. Two feeding cups.
28. Spittoon.
29. Tea-equipage.
30. Tea-kettle.
31. Medicine measure.
32. Apparatus for keeping food warm, with lamp.
33. Flowers.
34. A fan.
35. A Rimmel's perfume vaporiser.
36. Lucifers.

Before a room is occupied by a patient who has been operated on, it should be thoroughly cleaned ; the walls and ceiling should be well brushed, the carpet taken away and the floor well scrubbed with soda. All curtains and chintz furniture should be removed, old window-blinds replaced by new green ones, and the window made to open readily at the top and bottom. A fire or oil-lamp should be lighted in the fireplace to maintain a circulation of air. If the season require a fire, the iron fender should be removed and replaced by a wooden tray of sand or ashes, to prevent the noise of cinders and fire-irons falling on the fender and hearth. It is well also to flush all the drains, water-closets and sinks in the house with disinfecting fluid one or two days before the operation, and a store of Sir William Burnett's, or similar disinfecting fluid, should be made ready to neutralise the fœtid odours of discharges. The excreta of typhoid and other fever patients should be mixed with strong commercial hydrochloric acid before they are thrown into the house drains.

When possible, it is a great advantage to have two beds of similar height and size, that the patient may occupy them alternately. The cool bed refreshes the patient greatly, and the vacated bed is easily cleaned and aired without fatigue to the sick person.

SEDATIVES.

1. Tincture of opium.
2. Solution of morphia, and hypodermic syringe.
3. Morphia suppositories.
4. Ice.
5. Chloral-hydrate solution.
6. Solution of bromide of ammonia.

RESTORATIVES.

1. Brandy, champagne, sherry.
2. Eau-de-cologne.
3. Liquor ammoniæ.
4. Smelling salts, sal volatile.
5. Nitrite of amyl.
6. Ether.
7. An electrical battery.

8. Chicken-broth.
9. Beef-tea.

10. Milk.
11. Lime-water.
12. Soda-water.
13. Eggs.
14. French bread.
15. Biscuits.
16. Arrow-root.
17. Liebig's extract of meat.
18. Brand's extract of beef.

THE ARREST OF HÆMORRHAGE.

1. Tourniquet.
2. Artery forceps.
3. Torsion forceps.
4. Spencer Wells' forcipressure forceps (several pairs).
5. Tenaculum.
6. Hare-lip pins.
7. Acupressure needles.
8. Nævus needle.
9. Wire nippers.
10. Solution of perchloride of iron (equal parts of the salt and water).
11. Solid perchloride of iron.

12. Friar's balsam.
13. Ice, ice-cold water.
14. Paquelin's cautery.
15. Cautery irons.
16. Galvanic cautery wire.
17. Ligatures, silk, fine hemp, whipcord, and catgut.
18. Lint.
19. Amadou.
20. Bandages.
21. Compressed sponge.
22. Scissors.
23. Sutures.
24. Glass syringe.

(See List for the Ligature of Arteries.)

INSTRUMENTS AND APPLIANCES FOR THE FOLLOWING OPERATIONS ABOUT THE HEAD AND NECK.

TREPHINING THE SKULL.
OPERATIONS ON THE EYE.
HARE-LIP.
RESECTION OF THE JAW.
EXCISION OF THE TONGUE.

CLEFT PALATE.
EXCISION OF TONSILS.
LARYNGOTOMY.
TRACHEOTOMY.

[For requisites of Lister's antiseptic method, see p. 211.]

TREPHINING THE SKULL.

1. Scalpel.
2. Trephines—several crowns.
3. Hey's saw.
4. Elevator.
5. Stout dissecting forceps.
6. Brush to clean away the bone-dust.
7. Probe.
8. Quill, cut like a tooth-pick, to clear the groove of bone-dust.
9. Lenticular knife.
10. Small polypus forceps.
11. Sponges.
12. Lint.
13. Ice.
14. Bandages.
15. Macintosh sheet.

OPERATIONS ON THE EYE.

Strabismus.

1. Anesthetic, if used.
2. Lawrence's or other head-fixer.
3. Specula, of different sizes.
4. Toothed forceps.
5. Strabismus scissors and hooks.
6. Fine curved needles, and *finest* thread.
7. Lint.
8. Ice, or cold water.
9. Eye bandage.
10. Sponges.

Extirpation of the Eyeball.

1. Anesthetic.
2. Head-fixer.
3. Speculum.
4. Toothed forceps.
5. Curved scissors.
6. Strabismus hook.
7. Small and large sponges.
8. Dissecting forceps.
9. Ice-cold water.
10. Perchloride of iron.
11. Fine curved needles, and *finest* silk thread.
12. Lint.
13. Pulled lint.
14. Basin.
15. Bandage.
16. Sponges.

OPERATIONS ON THE EYE—*continued.*

Cataract (Congenital).

1. Atropine drops.
2. Anesthetic.
3. Head-fixer.
4. Speculum.
5. Lacerating needles.
6. Toothed forceps, to fix eyeball.
7. Lint.
8. Cold water.
9. Gelatine plaster.
10. Wool and bandage.
11. Sponges.

Cataract (Senile).

1. Atropine drops.
2. Anesthetic.
3. Head-fixer.
4. Speculum.
5. Lacerating needles.
6. Toothed forceps.
7. Cataract knives.
8. Scoop.
9. Platinum spatula, to adjust edges of wound.
10. Small sponges.
11. Basin and water.
12. Gelatine plaster.
13. Wool and bandage.
14. Lint.

Iridectomy.

1. Anesthetic, if used.
2. Head-fixer.
3. Specula.
4. Toothed forceps.
5. Right-angled knives.
6. Iris forceps.
7. Iris hooks.
8. Capsule scissors, if required.
9. Light curved scissors.
10. Lint.
11. Gelatine plaster.
12. Wool and bandage.
13. Atropine drops.
14. Small sponges and basin.
15. Hot and cold water.

HARE-LIP.

1. Scalpel.
2. Artery forceps, or sharp hook.
3. Hare-lip pins.
4. Wire nippers.
5. Dentist's silk twist.
6. Strapping plaster.
7. Silver suture.
8. Collodion.
9. Scissors.
10. Cheek compressor.
11. Sponges.
12. Large towel to wrap the child in.
13. Chloroform, if used.
14. Bone nippers.
15. Sequestrum forceps.
16. A knitting needle for checking deep hæmorrhage by actual cautery ; and
17. Spirit lamp to heat it.

RESECTION OF THE JAW, AND TUMOURS CONNECTED WITH IT.

1. Scalpels.
2. Artery forceps.
3. Torsion forceps.
4. Forcipressure forceps (several pairs).
5. Ligatures.
6. Retractors.
7. Tooth forceps.
8. Narrow saw.
9. Hey's saw.
10. Bone-cutting forceps.
11. Lion forceps.
12. Sequestrum forceps.
13. Gouges.
14. Chisel,
15. Gag.
16. Hare-lip pins.
17. Wire nippers.
18. Actual cautery.
19. Perchloride of iron.
20. Ice.
21. Sutures, silk and wire.
22. Solution of chloride of zinc.
23. Lint.
24. Bandages.
25. Plaster.
26. Collodion.
27. Macintosh sheet.
28. Small sponges tied on sticks.
29. Larger sponges.
30. Anesthetic and inhalér.

EXCISION OF THE TONGUE.

1. Scalpel.
2. Torsion forceps.
3. Artery forceps.
4. Vulsellum, or tongue forceps.
5. Forcipressure forceps (several pairs).
6. Gag.
7. Archimedian bone-drill.
8. Cheek retractor.
9. Incisor tooth forceps.
10. Narrow saw.
11. Nævus needle, and
12. Half a yard of thick whipcord.
13. Stout copper wire.
14. Key for twisting the wire tight.
15. Wire nippers.
16. Stout acupressure needle.
17. Hare-lip pins.
18. Galvanic écraseur.
19. Two écraseurs.
20. Curved scissors.
21. Needle for passing the chain.
22. Sharp and blunt hooks.
23. Stout silk.
24. Metallic sutures.
25. Ligatures.
26. Ice.
27. Perchloride of iron.
28. Cautery irons.
29. Solution of chloride of zinc.
30. Collodion.
31. Small sponges mounted on sticks ; larger sponges.
32. Brandy.
33. Anesthetic and inhaler.
34. Lint.
35. Macintosh sheet.
36. Syringe.
37. Laryngotomy tube, and Trendlenburg's bag.

1. Long-handled, narrow-bladed knife.
2. Fergusson's knife, for dividing the palatine muscles.
3. Long-handled curved scissors.
4. Slender toothed forceps, to seize the soft palate.
5. Long-handled needles, with the eye at the point for passing sutures.
6. Long-handled fine hooks, blunt and sharp.
7. Smith's gag.
8. Smith's elevator, for the hard palate.
9. Chisel for the hard palate.
10. Fine blue silk or catgut.
11. Long-handled ordinary straight scissors.
12. Ice, ice-cold water.
13. Macintosh sheet.
14. Sponges set on sticks.
15. Glass syringe for washing the mouth.
16. Chloroform and inhaler.
17. Bandage.

EXCISION OF TONSILS.

1. Long vulsellum.
2. Tonsil guillotines; or, probe-pointed bistoury, the posterior two-thirds of the blade covered with lint or plaster; or, tonsil excisors of various kinds.
3. Ice, ice-cold water.
4. Gag (for children).

To repress Hæmorrhage from the Tonsils.

1. A small lump of ice held in a long vulsellum against the tonsil.
2. Small sponge on the end of a stick, dipped in ice-cold solution of perchloride of iron, or other styptic.
3. Long straight polypus or gullet forceps with padded ends, to compress the tonsil; one blade being passed within the mouth, the other outside against the neck.
4. For the apparatus for deligating the common carotid artery, see p. 328.

LARYNGOTOMY IN THE CRICO-THYROID SPACE.

1. Scalpel.
2. Laryngotomy tube.
3. Artery forceps, and ligature silk.
4. Torsion forceps.
5. Tapes.
6. Sharp and blunt hooks.
7. Two hooked forceps.
8. Sponges.
9. Silk suture.

TRACHEOTOMY.

<table>
<tr><td>1. Scalpel.</td><td>10. Sponges.</td></tr>
<tr><td>2. Dissecting forceps.</td><td>11. If a foreign body is lodged in the windpipe, forceps of various kinds to extract with.</td></tr>
<tr><td>3. Blunt hooks.</td><td></td></tr>
<tr><td>4. Sharp hook to draw forward trachea.</td><td></td></tr>
<tr><td>5. Double trachea tube and tapes.</td><td>12. Probes and flexible No. 8 or 9 urethral bougie.</td></tr>
<tr><td>6. Trachea-dilator, to assist introduction of tube.</td><td>13. Strapping plaster.</td></tr>
<tr><td>7. Torsion forceps.</td><td>14. Silver suture.</td></tr>
<tr><td>8. Artery forceps.</td><td>15. Scissors.</td></tr>
<tr><td>9. Ligatures.</td><td>16. Macintosh sheet.</td></tr>
</table>

INSTRUMENTS, &c., FOR THE FOLLOWING OPERATIONS ABOUT THE TRUNK.

<table>
<tr><td>REMOVAL OF THE BREAST OR TUMOURS.</td><td>EXTIRPATION OF THE CERVIX UTERI.</td></tr>
<tr><td>NÆVUS.</td><td>AMPUTATION OF THE PENIS.</td></tr>
<tr><td>TAPPING THE PLEURA.</td><td>CIRCUMCISION.</td></tr>
<tr><td>TAPPING THE BELLY.</td><td>EXCISION OF TESTIS.</td></tr>
<tr><td>COLOTOMY.</td><td>TAPPING A HYDROCELE.</td></tr>
<tr><td>OVARIOTOMY.</td><td>VESICO-VAGINAL FISTULA.</td></tr>
<tr><td>CÆSARIAN SECTION.</td><td>RETENTION OF URINE.</td></tr>
<tr><td>STRANGULATED HERNIA.</td><td>EXTERNAL URETHROTOMY.</td></tr>
<tr><td>RADICAL CURE OF HERNIA.</td><td>LITHOTOMY.</td></tr>
<tr><td>HÆMORRHOIDS.</td><td>LITHOTRITY.</td></tr>
<tr><td>FISTULA IN ANO.</td><td>REMOVING FOREIGN BODIES FROM THE URETHRA AND BLADDER.</td></tr>
<tr><td>CLEFT PERINÆUM.</td><td></td></tr>
</table>

REMOVAL OF THE BREAST OR TUMOURS.

1. Scalpel.
2. Artery forceps.
3. Torsion forceps.
4. Dissecting forceps.
5. Forcipressure forceps (several pairs).
6. Double hook, or vulsellum.
7. Blunt and sharp hooks.
8. Tenaculum.
9. Fine ligatures.
10. Half-yard of stout whipcord, for irremovable glands.
11. Wire or silk sutures.
12. Lint.
13. Diachylon plaster.
14. Oiled silk.
15. Sponges.
16. Scissors.
17. Three-inch wide rollers.
18. Folded towel compress.
19. Anesthetic and inhaler.
20. Waterproof sheet.
21. Drainage tubes.

NÆVUS.

1. Scissors.
2. Lint.
3. Anesthetic and inhaler.
4. Sponge and water.
5. Collodion.

For Ligature.

1. Nævus needles, straight and curved.
2. Suture silk, stout compressed whipcord.
3. Scalpel.

For Cauterising.

1. Acupressure needles.
2. Spirit lamp.
3. Galvanic cautery.
4. Paquelin cautery.

For Excising.

1. Scalpel.
2. Dissecting forceps.
3. Toothed forceps.
4. Blunt hooks.
5. Artery forceps.
6. Sharp hooks.
7. Forcipressure forceps.
8. Ligatures.
9. Lint.
10. Bandage.
11. Sponges.
12. Sutures.
13. Diachylon plaster.

NÆVUS—*continued*.

For Injecting.

1. Pravaz's injecting syringe.
2. Solution of perchloride of iron.

For Passing Setons.

1. Small suture needles, or
2. Large handled needles.
3. Fine silk suture.
4. Coarse silk sutures, steeped in perchloride of iron.

TAPPING THE PLEURA.

1. The aspirator.
2. Small scalpel.
3. Trocar, fitted with india-rubber tube.
4. Bucket.
5. Lint.
6. Collodion.
7. Strapping.
8. Scissors.
9. Brandy.
10. Macintosh sheet.

TAPPING THE BELLY.

1. Body bandage (see p. 11).
2. Scalpel.
3. Trocar, with india-rubber tube.
4. Suture.
5. Hare-lip pins.
6. Wire nippers.
7. Silk ligature.
8. Collodion.
9. Scissors.
10. Lint.
11. Bucket.
12. Diachylon plaster.
13. Brandy.

COLOTOMY AND GASTROTOMY.

1. Scalpel.
2. Probe-pointed bistoury.
3. Directors.
4. Dissecting forceps.
5. Artery forceps.
6. Retractors.
7. Thick silk suture, threaded at each end through a curved needle.
8. Fine ligatures.
9. Sponges.
10. Lint.
11. Tow.
12. Oil.
13. Macintosh sheet.
14. Large syringe full of warm water.
15. Anesthetic and inhaler.
16. Higginson's pump for inflating the bowel.

OVARIOTOMY.

1. Room raised to temperature of 70° F.
2. Scalpel.
3. Probe-pointed bistoury.
4. Director.
5. Scissors, large and small.
6. Dissecting forceps.
7. Long slightly curved needles, two threaded on each suture.
8. Drainage tubes (glass or rubber).
9. Aneurism needles, threaded.
10. Torsion forceps.
11. Artery forceps.
12. Broad retractors.
13. Blunt hooks.
14. Hare-lip pins.
15. Wire nippers.
16. Forcipressure forceps.
17. Nélaton's ovariotomy forceps.
18. Vulsella.
19. Wire écraseur.
20. Needle holder.
21. Clamp.
22. Vulcanised india-rubber tubing, to fix on cannula.
23. Ovariotomy trocar.
24. Catgut ligatures.
25. Strong whipcord.
26. Silk sutures and silver wire.
27. Diachylon plaster.
28. Catheter for emptying the bladder.
29. Soft napkins.
30. Bandage or laced napkin.
31. Anesthetic and inhaler.
32. Cautery irons.
33. Cautery clamp.
34. Perchloride of iron.
35. Safety pins.
36. Sponges (counted).
37. Cotton wool, iodoform wool.
38. Warm flannels.
39. Brandy and ammonia.
40. Macintosh sheets.
41. Hypodermic syringe.
42. Bull's-eye lantern.
43. 2-ounce enema syringe.

CÆSARIAN SECTION.

1. Room maintained at temperature of 70° F.
2. Catheter to empty bladder.
3. Large scalpel.
4. Straight probed-pointed bistoury.
5. Director.
6. Large blunt hooks.
7. Vulsella.
8. Large syringe and vaginal tube, for washing out uterus per vaginam.
9. Artery forceps.
10. Dissecting forceps.
11. Ligatures, fine and stout whipcord.
12. Scissors.
13. Torsion forceps.
14. Forcipressure forceps.
15. Hare-lip pins, stout and long.
16. Suture silk.
17. Long straight or slightly curved needles, threaded two on each suture.
18. Fine sutures.
19. Folded linen compress.
20. Broad body roller.
21. Flannel.
22. Cotton wool, iodoform wool.
23. Warm flannels.
24. Anesthetic and inhaler.
25. Sponges (counted).
26. Macintosh sheets.
27. Hypodermic syringe.
28. 2-ounce enema syringe.

For Porro's Method, in addition.

1. Clamp.　　　　　　　　2. Kœberle's écraseur, or serre-nœud.

STRANGULATED HERNIA.

1. Scalpel.
2. Straight bistoury.
3. Probe-pointed bistoury.
4. Hernia knife.
5. Narrow director.
6. Broad director.
7. Dissecting forceps.
8. Blunt hooks.
9. Fine hook for a very tense sac.
10. Artery forceps.
11. Torsion forceps.
12. Probe.
13. Scissors.
14. Ligatures.
15. Sutures.
16. Lint.
17. Diachylon plaster.
18. Three-inch wide roller.
19. Compress.
20. Razor.
21. Sponges.
22. Half grain of morphia suppository.
23. Anesthetic and inhaler.
24. Macintosh sheet.
25. Oil.
26. Aspirator.

WOOD'S OPERATION FOR RADICAL CURE OF HERNIA.

1. Razor, or scalpel, for shaving the groin.
2. Tenotomy knife.
3. Needle.
4. Compressed whipcord, well waxed and soaped.
5. Glass or box-wood compress.
6. Two pads of lint.
7. Lint.
8. Bandage.
9. Scissors.
10. Sutures.
11. Collodion.
12. Sponges.
13. Anesthetic and inhaler.
14. Carbolic oil.
15. Macintosh sheet.

HÆMORRHOIDS.

External,—Excision of.

1. Vulsellum, or hook, or ringed forceps.
2. Knife-edged scissors curved on the flat.
3. Artery forceps.
4. Torsion forceps.
5. Ligatures.
6. Lint and T-bandage.
7. Smith's clamp.
8. Cautery iron.
9. Paquelin's cautery.
10. Sponges.
11. Macintosh sheet.
12. Opium suppository.
13. Nitric acid, fuming.
14. Wooden spatula.
15. Glass speculum.
16. Oil.
17. Anesthetic and inhaler.

HÆMORRHOIDS—*continued.*

Internal.

1. Enema of warm water.
2. Hook, vulsellum, or ring forceps.
3. Thin compressed whipcord.
4. Nævus needles threaded, to transfix the base of the pile.
5. Scissors.
6. Smith's clamp.
7. Ice.
8. Solid perchloride of iron.
9. Lint and cotton wool.
10. Opium suppository.
11. Oil.
12. Sponges.
13. Macintosh sheet.
14. Anesthetic and inhaler.

PROLAPSUS ANI.

The same as for hæmorrhoids.

FISTULA IN ANO.

1. Probes of various sizes, some grooved.
2. Director.
3. Probe-pointed curved bistoury.
4. Straight sharp-pointed bistoury.
5. Tenaculum.
6. Threaded curved needle set in a handle.
7. Torsion forceps.
8. Artery forceps.
9. Stout ligature.
10. Cautery iron.
11. Lint, cotton wool.
12. Sponges.
13. Compress.
14. T-bandage.
15. Oil.
16. Anesthetic and inhaler.
17. Macintosh sheet.
18. Suppository.

CLEFT PERINÆUM.

1. Scalpel.
2. Hooked forceps, or vulsellum.
3. Toothed forceps.
4. Curved nævus needles, threaded.
5. Stout whipcord, or stout silk.
6. Suture silk.
7. Wire suture.
8. Glass rods, or No. 12 bougies.
9. Collodion.
10. Lint.
11. Sponges on handles.
12. Oil.
13. Anesthetic and inhaler.
14. Catheter.

EXTIRPATION OF THE CERVIX UTERI.

1. Two hooked forceps, or long vulsella.
2. Specula (bivalve and duck-bill).
3. Ecraseur.
4. Two wool holders.
5. Small sponges on sticks.
6. Scalpel.
7. Straight probe-pointed bistoury.
8. Bistoury curved on the flat.
9. Long-handled straight scissors.
10. Long-handled curved scissors.
11. Artery forceps.
12. Ligatures.
13. Hare-lip pins.
14. Dentist's silk.
15. Cautery irons.
16. Perchloride of iron.
17. Solution of chloride of zinc.
18. Ice.
19. Lint (if plugging necessary, see p. 152).
20. Compressed sponge in pieces.
21. Soft silk handkerchief.
22. Cotton wool, iodoform wool.
23. Suppository.
24. Anesthetic and inhaler.

AMPUTATION OF THE PENIS.

1. Tape to tie round the root of penis, or Clover's disk.
2. Straight bistoury.
3. Scalpel.
4. Artery forceps.
5. Torsion forceps.
6. Dissecting forceps.
7. Scissors.
8. Fine ligatures.
9. Fine suture, to fix the flap of mucous membrane of the urethra.
10. Flexible catheter.
11. Lint.
12. Tape.
13. Ice.
14. Anesthetic and inhaler.
15. Carbolic oil.
16. Macintosh sheet.

CIRCUMCISION.

1. Small straight bistoury.
2. Polypus forceps.
3. Half-inch wide tape, or Clover's disk.
4. Artery forceps.
5. Fine ligatures.
6. Silk suture.
7. Torsion forceps.
8. Scissors.
9. Lint.
10. Ice.
11. Anesthetic and inhaler.
12. Macintosh sheet.
13. Otis' coil refrigerator.
14. Diachylon, or other plaster.

EXCISION OF TESTIS.

1. Scalpel, or bistoury.
2. Large sharp hook.
3. Blunt hooks.
4. Stout whipcord.
5. Fine ligatures.
6. Artery forceps.
7. Sutures, silk or wire.
8. Scissors.
9. Anesthetic and inhaler.
10. Lint.
11. Plaster.
12. Oiled silk.
13. Macintosh sheet.

TAPPING AND INJECTING A HYDROCELE.

1. Trocar.
2. Vulcanite syringe, with nozzle to fit cannula.
3. Solution of iodine.
4. Suspender for the testicles.
5. Lint.
6. Collodion.
7. Plaster.
8. Carbolic oil.

VESICO-VAGINAL FISTULA.

1. Duckbill speculum.
2. Scalpels set in long handles, curved blades.
3. Long-handled curved scissors.
4. Long-handled forceps.
5. Long-handled hooked forceps.
6. Long-handled sharp hook.
7. Narrow curved spatula.
8. Needles to carry silver wire suture.
9. Silver wire.
10. Probes.
11. Short curved needles, to carry fine silk suture.
12. Needle holder.
13. Suture tightener.
14. Lead clamp.
15. Awl for perforating the clamp.
16. Clamp adjuster.
17. Split shot.
18. Forceps for placing and pinching the shot.
19. Small sponges in holders.
20. Large sponges.
21. Sims' catheter.
22. Ice-cold water.
23. Syringe and flexible catheter for injecting the bladder, to test the perfect closure of the fissure.
24. Anesthetic and inhaler.
25. Oil.
26. Macintosh sheet.

RETENTION OF URINE.

1. Flexible catheters (Nos. 8 to ½), (English and French).
2. Filiform bougies (No. ¼).
3. Silver catheters
4. Long prostatic, and coudé catheters.
5. Carbolic oil (1 in 16).
6. Glass syringe.
7. Injecting bottle.
8. Tapes and string.
9. Strapping plaster.
10. Scissors.
11. Trocar for tapping per rectum.
12. Aspirator.
13. Tincture of opium.
14. Morphia suppository.
15. Anesthetic and inhaler.
16. Hot bath.

EXTERNAL URETHROTOMY.

1. Scalpels.
2. Catheters, silver.
3. Catheters, flexible.
4. Four feet of india-rubber tubing to fit catheter.
5. Probes, straight, and grooved with flat handles.
6. Syme's shouldered narrowed grooved staff ; or
7. Marshall's long jointed grooved sound, with flexible catheter sliding on it ; or
8. Wheelhouse's hooked staff.
9. Probe director and tapering gorget.
10. Lithotomy tapes or anklets.
11. Tapes.
12. Tenaculum ; two hooked forceps with catch.
13. Artery forceps.
14. Ligature.
15. Sponges.
16. Carbolic oil.
17. Ice.
18. Anesthetic and inhaler.
19. Needles threaded with blue silk.

INTERNAL URETHROTOMY.

1. Catheters, silver.
2. Catheters, flexible.
3. Urethrotomes.
4. Meatome.
5. Urethrometer.
6. Bullet sounds.
7. Sponges.
8. Carbolic oil.
9. Silk, or fine soft twine.
10. Macintosh sheet.
11. Anesthetic and inhaler.

LITHOTOMY.

1. Pair of lithotomy tapes or anklets.
2. Sound.
3. Staff.
4. Lithotomy scalpels.
5. Forceps.
6. Scoops.
7. Searcher.
8. Four-ounce syringe with a bulbous nozzle.
9. Bistouri caché.
10. Tubes with and without petticoats ; or Brown's air-tampon tubes.
11. Artery forceps.
12. Tenaculum.
13. Ligatures.
14. Lint.
15. Half grain morphia suppository.
16. Anesthetic and inhaler.
17. Carbolic oil.
18. Sponges.
19. Bandages and tape.
20. Gorget.
21. Lithotrite or other stone-crusher.

LITHOTRITY.

1. Lithotrites, fenestrated and flat.
2. Hollow sound, with a short beak.
3. Apparatus for washing out the crushings.
4. Carbolic oil.
5. Lithotrity catheters.
6. Urethral forceps.
7. Meatome.
8. Injecting bottle, with nozzle to fit the hollow sound.
9. Half grain morphia suppository.
10. Hot linseed poultice or fomentation.
11. Firm bolster or pillow.
12. Blotting paper, box, and straining muslin for collecting débris.
13. Anesthetic and inhaler.

FOR REMOVING FOREIGN BODIES FROM THE URETHRA AND BLADDER.

1. Catheters, silver, flexible.
2. Sounds of different curves.
3. Bougies à boule.
4. Urethral forceps with long blades.
5. Fine dressing forceps.
6. Fine polypus forceps.
7. Coxeter's urethral forceps.
8. Urethral lithotrite forceps.
9. Hunter's tube forceps.
10. Loop of wire set in long handle.
11. Leroy d'Etiolles' jointed scoop.
12. Leroy's tube and sliding hook.
13. Charrière's hair-pin retractor.
14. Charrière's bougie retractor.
15. Three-branched forceps.
16. Endoscope.
17. Carbolic oil.
18. Glass syringe.
19. Scalpel.
20. Dissecting forceps.
21. Hook.
22. Artery forceps.
23. Ligature.
24. Sutures.
25. Tapes.
26. Sponges.
27. Anesthetic and inhaler.

TRANSFUSION OF BLOOD.

See PAGE 174.

INSTRUMENTS FOR OPERATIONS ON THE LIMBS.

LIGATURE OF THE LARGER ARTE-
RIES.
RESECTIONS OF THE HEAD OF
THE HUMERUS, ELBOW, HIP,
AND KNEE.
REMOVAL OF NECROSED BONE.

AMPUTATIONS AT THE SHOULDER-
JOINT ; ARM ; FOREARM ; AND
WRIST ; METACARPUS ; HIP ;
THIGH ; AND LEG ; BY SYME'S
AND CHOPART'S OPERATIONS ;
METATARSUS.

LIGATURE OF THE LARGE ARTERIES.

1. Scalpel.
2. Grooved director.
3. Dissecting forceps.
4. Broad grooved director.
5. Probe.
6. Blunt hooks.
7. Metallic retractors.
8. Aneurism needle.
9. Ditto, helix curve.
10. Waxed compressed whipcord.
11. Tourniquet.

12. Artery forceps.
13. Fine ligatures.
14. Sutures.
15 Scissors.
16. Sponges on sticks.
17. Larger sponges.
18. Strapping.
19. Lint.
20. Bandage.
21. Anesthetic and inhaler.
22. Macintosh sheet.

RESECTION OF JOINTS.

1. Bistouries.
2. Saw (Irish bow-saw).
3. Retractors.
4. Lion forceps.
5. Blunt hooks.
6. Artery forceps.
7. Torsion forceps.
8. Tenacula.
9. Linen retractors.
10. Ligatures.
11. Sutures.
12. Cautery iron.
13. Wire nippers.
14. Lint.
15. Bandages.
16. Strapping.

17. Oiled silk.
18. Sponges.
19. Macintosh sheet.
20. Anesthetic and inhaler.
21. Drainage tubes.
22. Tourniquet ; or Esmarch's
elastic band.
23. Key to compress the subclavian artery, if the
shoulder.
24. If the elbow, angular splint.
25. If the head of the femur, a
straight bracketed splint.
26. If the knee, back splint and
pads.
27. If the ankle, moulded splint.

REMOVING NECROSED BONE.

1. Scalpels.
2. Straight and curved bistouries with sharp and probe points.
3. Long and short probes.
4. Directors.
5. Retractors.
6. Volkmann's sharp spoons.
7. Chisels, and blunt periosteum elevators.
8. Bone forceps (various).
9. Gouge forceps.
10. Sequestrum forceps.
11. Lion forceps.
12. Osteotrites.
13. Trephine.
14. Gouges.
15. Artery forceps.
16. Tenaculum.
17. Torsion forceps.
18. Ligatures.
19. Sutures.
20. Lint and cotton wool.
21. Sponges.
22. Bandages.
23. Oiled silk.
24. Macintosh sheet.
25. Anesthetic and inhaler.
26. Tourniquet; or, Esmarch's elastic band.
27. Cautery iron.
28. Drainage tubes.

AMPUTATION AT THE SHOULDER JOINT.

1. Long amputating knife.
2. Artery forceps.
3. Fine ligatures, stout ligatures.
4. Tenaculum.
5. Lion forceps.
6. Forcipressure forceps.
7. Key · to compress subclavian artery.
8. Sutures.
9. Diachylon plaster.
10. Bandages.
11. Lint and cotton wool.
12. Oiled silk.
13. Sponges.
14. Cautery irons.
15. Ice.
16. Macintosh sheet.
17. Anesthetic and inhaler.
18. Drainage tubes.

AMPUTATION OF THE ARM.

1. Tourniquet; or, Esmarch's elastic band.
2. Amputating knife.
3. Saw.
4. Artery forceps.
5. Torsion forceps.
6. Ligatures.
7. Sutures.
8. Tenaculum.
9. Bone nippers.
10. Diachylon plaster
11. Lint.
12. Oiled silk.
13. Bandages.
14. Wadding.
15. Straight splints.
16. Sponges.
17. Macintosh sheet.
18. Anesthetic and inhaler.
19. Cautery iron.
20. Drainage tubes.

If a circular amputation—

21. Split linen retractor; and
22. Round-pointed straight-edged knife.

AMPUTATION OF THE FOREARM AND WRIST.

1. Tourniquet ; or,
    Esmarch's elastic band.
2. Large bistoury, or small amputating knife.
3. Torsion forceps.
4. Artery forceps.
5. Saw.
6. Bone nippers.
7. Tenaculum.
8. Fine ligatures.
9. Sutures.
10. Lint.
11. Wool.
12. Splints.
13. Bandages.
14. Oiled silk.
15. Diachylon plaster.
16. Sponges.
17. Macintosh sheet.
18. Anesthetic and inhaler.
19. Drainage tubes.

AMPUTATION OF METACARPAL BONES AND PHALANGES.

1. Tourniquet ; or,
    Esmarch's elastic band.
2. Narrow-bladed bistoury.
3. Bone nippers.
4. Lion forceps.
5. Torsion forceps.
6. Artery forceps.
7. Fine ligatures.
8. Small sutures.
9. Diachylon plaster.
10. Narrow bandage.
11. Wadding.
12. Splint to support the arm and hand.
13. Bandage.
14. Lint.
15. Oiled silk.
16. Macintosh sheet.
17. Sponges.
18. Anesthetic and inhaler.

AMPUTATION AT THE HIP JOINT.

1. Lister's aorta compressor ; or, Esmarch's elastic band ; or, Davy's lever.
2. Long hip knife.
3. Scalpel.
4. Artery forceps.
5. Forcipressure forceps.
6. Stout and fine ligatures.
7. Torsion forceps.
8. Tenaculum.
9. A bone-holder or lion forceps, if the bone is too short to hold by the hand.
10. Silk and silver sutures.
11. Lint.
12. Diachylon plaster.
13. Sponges.
14. Bandages.
15. Cotton wool.
16. Ice.
17. Anesthetic and inhaler.
18. Macintosh sheet.
19. Drainage tubes.
20. Gouges, for carious acetabulum.
21. Periosteum elevators.

AMPUTATION OF THE THIGH AND LEG.

1. Tourniquet ; or,
     Esmarch's elastic band.
2. Amputating knife.
3. Scalpel.
4. Bone nippers.
5. Saw.
6. Torsion forceps.
7. Artery forceps.
8. Stout and fine ligatures.
9. Sutures.
10. Tenaculum.
11. Diachylon plaster.
12. Lint.
13. Cotton wool.
14. Bandages.
15. Oiled silk.
16. Cautery iron.
17. Ice.
18. Sponges.
19. Macintosh sheet.
20. Anesthetic and inhaler.
21. Drainage tubes.

In a circular amputation—

22. A round pointed knife.
23. A linen retractor split into two tongues for the thigh, and three for the leg, the centre one being well waxed.

AMPUTATION AT THE ANKLE AND FOOT.

(Syme's and Chopart's Operations.)

1. Tourniquet ; or,
     Esmarch's elastic band.
2. Strong bistoury.
3. Strong scalpel (Syme's).
4. Saw.
5. Lion forceps.
6. Bone nippers.
7. Artery forceps.
8. Torsion forceps.
9. Tenaculum.
10. Cautery iron.
11. Diachylon plaster.
12. Sutures.
13. Lint.
14. Bandages.
15. Oiled silk.
16. Wool.
17. Sponges.
18. Anesthetic and inhaler.
19. Macintosh sheet.
20. Drainage tubes.

AMPUTATION OF THE METATARSAL BONES AND TOES.

1. Tourniquet ; or,
     Esmarch's elastic band.
2. Straight bistoury.
3. Lion, or sequestrum forceps.
4. Narrow saw.
5. Bone nippers.
6. Tenaculum.
7. Torsion forceps.
8. Artery forceps.
9. Ligatures.
10. Sutures.
11. Cautery iron.
12. Diachylon plaster.
13. Lint and wool.
14. Bandages.
15. Macintosh sheet.
16. Anesthetic and inhaler.

INDEX.

A.

ABDOMEN, landmarks of, 266.
—— regions of, 270.
Abdominal tumours, 269.
Abscess of axilla, 261.
—— of spine, 263.
—— near wrist, 294.
Air-bath, the hot, 188.
Air-cushions, 170.
Amputations, 329, 330, 331.
Angular splint, 49.
—— union, 82.
Ankle, bony landmarks of, 305.
—— strapping the, 29.
Anterior angular ligament, 293.
—— jugular vein, 251.
—— temporal artery, guide to, 242.
—— tibial artery, 304.
Antiseptic dressing, 206.
Anus, 280.
—— parts felt within, 281.
Aorta, abdominal, 274.
—— arch of, 258.
Arch, palmar, wound of, 18.
Arm, landmarks of, 286.
Arnott's floating bed, 168.
Arteries, ligature of, 328.
Artificial nipples, 152.
—— respiration, 228.
Ascites, bandage for, 11.
Aspirator, the, 189.
Astragalus, head of, 305.
Auditory canal, 243.

Axilla, borders of, 255.
Axillary artery, 262, 283.
—— region, 261.

B.

BANDAGES, 1.
—— arm, 17.
—— belly, 11.
—— bleeding jugular vein, 8.
—— breast, 9.
—— capelline, 6.
—— eighteen-tail, 24.
—— elbow, 17.
—— fingers, 15.
—— foot, 22.
—— four-tail, 8.
—— groin, 10.
—— hand, 16.
—— head, 4.
—— heel, 22.
—— knee, 23.
—— knotted, 5.
—— leg, 22.
—— many-tail, 24.
—— plaster of paris, 92.
—— Croft's, 95.
—— Sayre's, 98.
—— Scultetus, 24.
—— shoulder, 74.
—— spica, 10, 15, 17.
—— starch, 86.
—— stump, 23.
—— T, 12.

Bandages, thumb, 15.
——— toe, 23.
——— winder, 2.
Band box-cradle, 103.
Basilic veins, 289.
Bath, the hot-air, 188.
——— the vapour, 189.
Bed, the floating, 168.
——— the sick, 312.
Bedsores, 168.
Belloc's sound, 143.
Belly, tapping the, 11, 320.
Bichloride of methylene, 227.
Bladder, relations of, 280, 281.
——— washing out the, 164.
Bleeding, cupping, 190.
——— at elbow, 19, 289.
——— jugular vein, at, 9.
——— leech bites, 193.
——— socket of tooth, 149.
——— tape, 19.
Blisters, 203.
Blood, transfusion of, 174.
Bloodless operations, 181.
Boracic acid dressing, 217.
——— ointments, 217.
Bougies, 156.
Brachial plexus, 264.
Bread poultice, 204.
——— cold, 205.
Breast, bandaging the, 9.
——— strapping the, 25.
——— removing, 319.
Brown's tampon, 167.
Bryant's rule, 297.
Bulbous ended catheter, 162.
Buttock, surface marks of, 295.
Button suture, 215.

C.

Cæsarian section, 321.
Calcaneum, separation of the epi-
physis of, 61.
Calf, surface marks of, 303.
Calomel fumigation, 185.
Cannulæ, Southey's, 12.
Cap for shoulder, 54.

Carbolic acid dressing, 206.
Carbolised catgut, 210.
Carotid artery, 251, 257.
Carr's splints, 41.
Casting in plaster of paris, 133.
Catheters, 154, 155, 156, 161, 162.
——— passing, 158.
——— ——— in the female, 163.
——— tying in, 165, 166.
Caustics, 202.
Cauteries, 200.
Cautery irons, 200.
——— galvanic, 201.
——— Paquelin's, 201.
Cerebrum, level of, 241.
Cervical plexus, 264.
Chassaignac's drainage tube, 194.
Chloroform, administration of, 222.
——— to the uterus, 234.
Chopart's amputation, landmarks for,
307.
Circumcision, 324.
Clavicle as a landmark, 252, 253, 255.
——— Sayre's strapping, 59.
——— ring pad for, 59.
——— figure of 8 for, 60.
Cleft palate, 317.
——— perinæum, 323.
Clove hitch, the, 113.
Clover's inhaler, 226.
Cœliac axis, position of, 275.
Coin-catcher, 170.
Cold injection, 139.
——— to head, 8.
——— irrigation, 18, 219.
Collodion, 237.
Colon, position of, 272.
Colotomy, 320
Common carotid artery, 251—257.
Compress, graduated, 5, 18.
Congelation by ether, 233.
Continuous extension, 82, 86.
Coronoid process, 247.
Corrigan's hammer, 202.
Coxeter's elastic perineal band, 77.
Cradles, 103, 104.
Cranial region, 241.
Cricoid cartilage, 250.
Crico-thyroid membrane, 250.

Croft's splints, 95.
Cupping, bleeding, 192.
—— dry, 190.
Cushions, water, 169.

D.

DAVEY's lever, 183.
Delirious patients, manacles for, 12.
Diaphragm, attachment of, 260.
Digastric muscle, 249.
Disjunction of articular surfaces, 47, 288.
Dislocations, 109.
       —— clavicle, 110.
       —— elbow, 115.
       —— distinction from fracture of, 46.
       —— fingers, 117.
       —— foot, 126.
       —— jaw, 109.
       —— hip, 118.
       —— knee, 125.
       —— patella, 125.
       —— radius, 117.
       —— shoulder, 111.
       —— thumb, 117.
Distribution of nerves, 264.
Domett's flannel roller, 1.
Dorsal artery of foot, 306.
   —— region of thorax, 262.
Double inclines, 83.
       —— slung, 84.
Drainage tubes, 194, 209.
Drawing teeth, 144.
Dry cupping, 190.
Dry heat, 206.
Drying starch bandages, 91.
Ductus ad nasum, 245.
Duodenum, 273.
Dupuytren's splint, 61.

E.

EARS syringing the 138.
Eight figure of 4

Elastic extension, 78.
    —— perineal band, 77.
    —— socks, 25.
    —— stirrup, 78.
Elbow, landmarks of, 286—290.
Elevator, 148.
Epididymis, 277.
Epigastric artery, 276.
Epiphysis of os calcis, separation of the, 61.
Epistaxis, 139.
Esmarch's elastic bandage, 181.
    —— irrigator, 221.
Essentials of a truss, 195.
Ether, administration of, 225.
   —— with chloroform, 227.
   —— spray, 233.
Ethidene dichloride, 226.
Extension in fractures, 85.
    —— in hip disease, 86.
    —— in the long splint, 78.
    —— by weight and pulley, 86.
External abdominal ring, 276.
    —— auditory passage, 243.
    —— carotid artery, 244.
    —— jugular vein, 253.
    —— malleolus, 305.
    —— plantar artery, 306.
Extirpation of cervix uteri, 324.
Eye douche, 137.
—— operations on, 314.

F.

FACE, landmarks of, 243.
Facial artery, 244.
   —— nerve, 245.
Felt jacket, 103.
Feeding by the stomach pump, 171.
Female, passing catheter for, 163.
Femoral artery, 298.
    —— hernia, 277.
Femur, condyles of, 299.
Fibula, head of, 299.
Fifth metatarsal bone as a landmark 307.
Figure of 8 turn, 4.

Fingers, surface marks of, 2, 95.
Fistula in ano, 323.
Flannel rollers, 1, 92.
Floating bed, 168.
Fomentations, 205.
Fontanelles, 228, 242.
Foot, landmarks of, 306.
Forceps, tooth, 144.
Forearm, landmarks of, 289—292.
Four-tail bandage, 8, 33.
Fractures, 32.
—— acromion, 57.
—— clavicle, 57.
—— Colles', 38.
—— near elbow, 46.
—— femur, shaft of, 74.
—— fibula, 61.
—— forearm, 42.
—— humerus, neck of, 53.
—— —— shaft of, 51.
—— —— lower end of, 46.
—— —— great tuberosity of, 56.
—— jaw, 32.
—— metacarpal bones, 36.
—— olecranon, 44.
—— os calcis, 61.
—— patella, 70.
—— pelvis, 36.
—— phalanges, 37.
—— radius, 38, 44.
—— rib, 35.
—— tibia, 62.
—— transverse ditto, 67.
—— ulna, 42.
Fumigation, 183.
—— local, 187.

G.

Gag, for stomach-pump, 171.
Gall bladder, position of, 272.
Gas and ether, 225.
Gastrotomy, 320.
Gauntlet for fractured radius, 40.
Gauze, antiseptic, 208.

General rules, 1.
Glove for metacarpal bones, 36.
Gluteal artery, 297.
Gooch's splints, 107.
Graduated compress, 5, 18.
Great trochanter as a landmark, 295.
Groin, spica for, 10.
—— landmarks of, 295.
Gum and chalk, 97.
Gustatory nerve, guide for cutting, 247.
Gutta-percha, 32, 36, 41, 46, 49, 53.

H.

HÆMORRHAGE, arrest of, 313.
—— after lithotomy, 167.
—— from socket, 149.
—— from the tonsils, 317.
Hæmorrhoids, 322.
Ham, 301.
Hamilton's splint, for children, 81.
Handle for extending the thumb, 117.
Hare-lip, 315.
Head, cold to, 8.
—— ice bladder for, 8.
—— landmarks of, 241.
Heart, area of, 257.
Hernia, strangulated, 322.
—— umbilical, 270.
Hip-disease, extension in, 85.
—— splint, 105, 127.
—— surface marks of, 295.
Hollow in front of elbow, 289.
Horizontal position in fractured clavicle, 57.
Horseshoe splint, 69.
—— tourniquet, 183.
Hot air bath, 188.
—— fomentations, 206.
—— water cushions, 206.
How to hold a roller, 2.
Humerus, fracture of, 46, 51, 53, 56.
—— junction of epiphysis, 285, 288.
Hyoid bone, 249.

I.

Ice bladders on head, 8.
—— cold injections, 139.
Ileum, position of, 273.
Iliac abscess, 295.
—— arteries, 275.
Improvised tourniquet, 181.
Incline, double, 83.
—— —— slung, 84.
India-rubber coil, 220.
Inguinal hernia, 275.
Injecting syringe, 153.
—— the urethra, 153.
Injection, subcutaneous, 235.
Innominate vessels, 56.
Instruments, lists of, 309.
Interdental splints, 34.
Internal abdominal ring, 276.
—— carotid artery, 252.
—— jugular vein, 257.
—— malleolus, 305.
—— mammary artery, 261.
—— saphenous vein, position of, 304.
Iodine blisters, 203.
Iodoform, 219.
Irons, cautery, 200.
Irrigation, 219.
Irrigator, Esmarch's, 221.
—— by spiral tube, 220.
Irritants, 202.
Ischiatic artery, 298.
Issues, 195.
Isthmus thyroideæ, 250.

J.

Jacket, strait, 12.
—— felt, 103.
—— plaster, 98.
Jaw, resection of, 316.
Jejunum, position of, 273.
Joints, strapping, 28.
Jugular vein, bandage for, 8.
Jury-mast, 102.

K.

Kidney, position of, 274.
Kite's tail plug, 153,
Knee, surface marks, 299.

L.

L-shaped splint for arm, 49.
Lachrymal sac, guide to, 244.
Landmarks, 242.
Large intestine, position of, 272.
Laryngoscope, 149.
Laryngotomy, 317.
Lateral region of neck, 252.
—— sinus, 242.
—— splints for the tibia, 67.
Latissimus dorsi, 263.
Leather splints, 49, 104.
Leeches, 192.
Leg, surface marks of, 303.
Leiter's tubes, 8, 221.
Ligature of arteries, 328.
Linea alba, 268.
—— operations in, 270.
Lingual artery, 249.
Lister's button suture, 216.
—— dressing, 206.
—— tourniquet, 183.
Liston's long splint, 74.
Lists of instruments, 309.
Lithotomy, 326.
—— position, 166.
Lithotrity, 327.
Liver, position of, 271.
Local anæsthesia, 233.
Long splint, 74.
Longitudinal sinus, 242.
Lower extremity, surface marks of, 295.
Lungs, area of, 259.
Lymphatic glands of neck, 254.
—— elbow, 288.

M.

McIntyre's splint, 63.
Manacles for delirous patients, 13.
Many-tail bandage, 24.
Materials for rollers, 1.
Membrana tympani, 243.
Mercurial fumigation, 183.
—— ointment, 30.
Metatarsal bone, resecting, 307.
Methylene, bichloride of, 227.
Middle meningeal artery, guide to, 242.
Millboard for splints, 87.
Mouth, cavity of, 246.
—— parts to be felt in, 247.
—— parts to be seen in, 246.

N.

Nævus, 319.
Nape of neck, 254.
Nares, plugging the, 140.
Nasal douche, 139.
Neck, surface guides of, 248.
Nelaton's line, 296.
Nipple, position of, 255,
—— shields, 152.
Nitrous oxyde, 227.
Nose, parts to be seen in, 245.

O.

Occipital artery, 243.
Olecranon, relations of, 287.
Omentum, position of, 273.
Omo-hyoid muscle, 253.
Ophthalmoscopy, 135.
Operating room, the, 311.
Os uteri, leeches to, 192.
Ovariotomy, 321.

P.

Palm, surface marks of, 294.

Palmar arch, wound of, 18.
Pancreas, position of, 273.
Paraffin, 97.
Parotid duct, guide to, 245.
—— gland, 245.
Passing catheters, 158.
Pasteboard splints, 88.
Patella, starch bandage for, 73.
—— back splint for, 72.
—— as a landmark, 299.
Penis, amputation of, 324.
Perineal band, 75.
Perinæum, landmarks of, 279.
Peronei, tendons of, 305.
Pharynx, parts to be felt in, 248.
Pirogoff's amputation, landmarks for, 307.
Pistol splint, 39.
Plaster of paris, bandage, 92.
—— splints, 95.
—— jackets, 98.
—— casting in, 132.
Pleura, tapping the, 262, 320.
Plugging nares, 140.
—— vagina, 152.
Points for vaccine, 237.
Polypi, starting point of, 246.
Pomum adami, 249.
Popliteal space, surface marks of, 301.
—— artery, 301.
—— tumours of, 301.
Poroplastic felt, 103, 108.
Position of bandager, 1.
—— for lithotomy, 166.
Posterior tibial artery, 304.
Poultice, 203.
—— bread, 204.
—— linseed, 204.
—— mustard, 202.
—— starch, 205.
Poupart's ligament, 268.
Prolapsus ani, 322,
Prostate, relations of, 281.
Protective, 209.
Psoas abscess, 295.
Pterygo-maxillary ligament, 247.
Pubes, as a landmark, 267, 276
Pudic artery, position of, 298.

Pulley, extension by weight, 86.
Pulleys, for dislocation, 119.
Pulverised fluids, 233.
Pump, the stomach, 171.
Pylorus, position of, 271.

R.

RADIAL artery at wrist, 291.
Ranine vessels, 247.
Removing a starch bandage, 92.
—— a plaster of paris bandage, 95.
—— necrosed bone, 330.
—— bodies from urethra, 327.
Resections, 328.
Respiration, artificial, 228.
Restoratives, 313.
Retention of urine, 325.
Reverse turn, 3.
Rib, roller for, 35.
—— strapping, 35.
Richardson's spray producer, 233.
Ring pad, 59.
Rollers, how to hold, 2.
—— varieties of, 1.
Roussel's transfusor, 177.
Rules, general, 1.
Ruptured tendo achillis, 61.

S.

SALICYLIC acid dressing, 218.
Salter's cradle, 65.
Sand bags, 97.
Saphena vein, 303.
Saphenous opening, 277.
Sayre's plaster jacket, 98.
—— strapping for broken collarbone, 59.
Scarpa's shoes, 126.
Sciatic nerve, 297.
Scotch sheet for long splint, 79.
Scott's bandage, 30.
Scrotum, contents of, 277.

Scultetus, bandage of, 24.
Second costal cartilage, 25.
Sedatives, 313.
Self-suspension, 99.
Separation of epiphysis of the os calcis, 61.
Separation of lower end of humerus, 47.
Septum narium, 245.
Serratus magnus, 256.
Setons, 194.
Shawl cap, 7.
Shot-bag weight, 86.
Shoulder-cap, 53.
—— surface-marks of, 283.
Sick room, the, 312.
Silicate of soda, 97.
Skull, thickness of, 42.
Slung double incline, 84.
Small intestine, 273.
Sock, elastic, 25.
Socket, a bleeding, 149.
Solar plexus, 275.
Sore nipples, 152.
Sound, Belloc's, 143.
Sounds, 156.
Spermatic cord, 277.
Spica for groin, 10.
—— shoulder, 17.
—— for thumb, 15.
Spinal nerves, origin of, 263.
Spines, as landmarks, 262.
Spiral cord, 220.
—— turn, 3.
Spleen, position of, 273.
Splint, the long, 74.
—— the hip, 105.
Spongio piline, 205.
Spongy bones, 231.
Spray producer, 210, 234.
Starch bandage, 87.
—— —— for patella, 73.
Sternal notch, 255.
Sterno-clavicular joint, 255.
Sterno-mastoid muscle, 251.
Sternum, as a landmark, 255.
—— first joint of, 255.
Stirrup extension, 77.
Stirrup extension, Brown's, 79.

Stockings, elastic, 25.
Stocking-web rollers, 1.
Stomach, position of, 271.
——— pump, the, 171.
——— washing the, 173.
Strait jacket, 12.
Strapping, 26.
——— ankle, 29.
——— the breast, 26.
——— joints, 28.
——— ——— with mercurial oint-
ment, 30.
——— ribs, 35.
——— testes, 27.
——— ulcers, 28.
Strictures of the rectum, 282.
——— urethra, 160.
Stromeyer's cushion, 50.
Stump, extending, 24.
Stumps, extracting, 148.
Subclavian artery, 252, 256.
——— vein, 256.
Subcutaneous injection, 235.
Submaxillary gland, 249.
Superficial temporal artery, 243.
Superior mesenteric artery, 273.
Superior thyroid artery, 249.
Supra-orbital artery, guide to, 243.
——— notch, 244.
Surface guides, 241.
Suspending the testes, 14.
Syme's amputation, landmarks for,
307.
Syringing the ears, 138.

T.

Taxis, the, 278.
T-bandage, 12.
Taking off a starch bandage, 91.
——— plaster of paris, 95.
Tape for bleeding, 19.
Tapping, bandage for, 11.
——— the chest, guides for, 262.
——— hydrocele, 325.
Temporal artery, 243.
Tendo achillis, ruptured, 60.

Tenotomy in club foot, 304.
Tents, 94.
Testes, strapping, 27.
——— suspending, 14.
Testing a truss, 196.
Testis, excision of, 325.
——— relations of, 277.
Tetanus, to feed in, 247.
Thomas's splints, 127.
Thorax, surface guides of, 255.
Thyro-hyoid membrane, 249.
Thyroid tumours, 251.
Tibia, tuberosities of, 299.
Tibialis anticus, tendon of, 335.
——— posticus, tendon of, 304.
Tongue, excision of, 316.
Tongue-tied, to cut, 247.
Tonsils, 246.
——— bleeding, 317,
——— excision of, 317.
——— mode of lancing, 246.
Tooth drawing, 144.
Tourniquets, 180.
——— Carte's, 183.
——— Esmarch's, 181.
——— Lister's, 183.
——— makeshift, 181.
——— Petit's, 180.
——— Signoroni's, 183.
Trachea, 249, 256.
Tracheotomy, 250, 318.
Transfusion of blood, 174.
Trapesius muscle, 263.
Trephining, 314.
Trusses, 195.
——— measuring for, 193.
——— femoral, 197.
——— inguinal, 197.
——— umbilical, 199.
——— Salmon and Ody's, 199.
Tubercle of scaphoid, 306.
——— os calcis, 306.
Tuberosity of upper jaw, 247.
Tubes, drainage, 194, 209.
Tumours of neck, 254.
——— removal of, 319.
Tying for lithotomy, 166.
——— in a flexible catheter, 166.
——— in a silver catheter, 165.

U.

ULNAR artery at wrist, 293.
Union, angular, 82.
Upper extremity, landmarks of, 283.
Urethra, conformation of, 158.
———— injecting the, 153.
———— position of, 280.
Urethral syringe, 153.
Urethrotomy, 326.
Urinary bladder, 269.
Urine, retention of, 325.
Uterus, chloroform to, 234.
———— enlargement of, 269.

V.

VACCINATION, 237.
Vagina, plugging the, 152.
Vapour bath, 189.
Varieties of turns, 3.
———— of splints for the elbow, 47.
Veins of hand, 291.
———— jugular, 251, 252, 253.
———— of leg, 304.
Vena cava, 257.
Venæsection, 19.

Vesicants, 202.
Vesicular seminales, position of, 282.
Vesico-vaginal fistula, 325.
Vienna paste, 202.
Volkman's spoon, 215.

W.

WAISTCOAT, strait, 12.
Washing out the bladder, 164.
———— stomach, 173.
Water-bed, 168.
Water-can weight, 86.
———— cushion, 169.
Weight for pulley, 86.
Wind-pipe, position of, 249, 256.
Wire gauze splints, 48.
Wound of palmar arch, 18.
———— opening in splint for, 94.
Wrist, landmarks of, 290, 293.

Z.

ZINC, chloride of, 215.
Zygoma, 241.

THE END.